CW01080672

A Managerial Philosophy of Technology

A Managerial Philosophy of Technology

Technology and Humanity in Symbiosis

Geoff Crocker
Technology Market Strategies

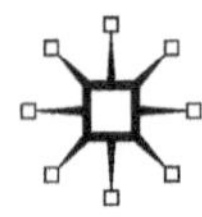

First published 2012 by
PALGRAVE MACMILLAN

Palgrave Macmillan in the UK is an imprint of Macmillan Publishers Limited, registered in England, company number 785998, of Houndmills, Basingstoke, Hampshire RG21 6XS.

Palgrave Macmillan in the US is a division of St Martin's Press LLC, 175 Fifth Avenue, New York, NY 10010.

Palgrave Macmillan is the global academic imprint of the above companies and has companies and representatives throughout the world.

Palgrave® and Macmillan® are registered trademarks in the United States, the United Kingdom, Europe and other countries

ISBN: 978–0–230–38914–4

This book is printed on paper suitable for recycling and made from fully managed and sustained forest sources. Logging, pulping and manufacturing processes are expected to conform to the environmental regulations of the country of origin.

A catalogue record for this book is available from the British Library.

A catalog record for this book is available from the Library of Congress.

10 9 8 7 6 5 4 3 2 1
21 20 19 18 17 16 15 14 13 12

Printed and bound in Great Britain by
CPI Antony Rowe, Chippenham and Eastbourne

Contents

List of Tables	viii
List of Figures	ix
Acknowledgements	x
About the Author	xi
1 Introduction	**1**
1.1 The structure of a story, theory, philosophy and management of technology	5
2 The Academic Literature	**6**
2.1 Definitions of technology	8
2.1.1 Rationality	8
2.1.2 Human enhancement	12
2.1.3 Cyborg ontology	14
2.1.4 Instrumentalism	15
2.2 Analytics of technology	19
2.2.1 Determinism	19
2.2.2 Autonomy	25
2.2.3 Social constructivism	27
2.2.4 Technocracy	28
2.2.5 Utopia and dystopia	29
2.2.5.1 Heidegger	29
2.2.5.2 Herbert Marcuse	32
2.2.5.3 Hans Jonas	33
2.2.5.4 Alfred Borgmann	34
2.2.6 New conceptualisations of technology – American empiricist phenomenology	34
2.3 Moderating technology – taming the beast?	37
2.3.1 Heidegger's 'saving power'	37
2.3.2 Ellul and Borgmann – faith and focal things	38
2.3.3 Habermas and Feenberg – democratisation	38
3 A Comprehensive Systems Network Philosophy of Technology	**41**
3.1 The model's assumptions	44

3.2 The model's entities 52
3.2.1 Nature 52
3.2.1.1 Time, force and field in nature 54
3.2.1.2 Probability in nature 55
3.2.1.3 Mathematics in nature 57
3.2.1.4 Purpose in nature 59
3.2.1.5 Infinity in nature 59
3.2.1.6 The nature of nature 60
3.2.2 Science 61
3.2.2.1 Science or not science – Karl Popper 64
3.2.2.2 Science – theory or paradigm? Thomas Kuhn 67
3.2.2.3 Philosophy of science 75
3.2.2.4 Induction 75
3.2.2.5 Reduction 78
3.2.2.6 Realism 79
3.2.3 Technology 81
3.2.3.1 A technology narrative 83
3.2.3.2 Textile technologies 88
3.2.3.3 Agricultural technologies 90
3.2.3.4 Propulsion technologies 93
3.2.3.5 Fuel cell technology 98
3.2.3.6 Medical technologies 100
3.2.3.7 Systems technology 104
3.2.3.8 Ecological technologies 108
3.2.4 Productivity 112
3.2.5 The economy 115
3.2.5.1 Forms of economy – command and market economies 115
3.2.5.2 Productivity in the real economy 124
3.2.6 Society 130
3.2.7 Ecology 144

4 Resolving and Managing the Model 146
4.1 The behavioural economics view of technology – innovation studies 146
4.2 Models of the management of technology in a market economy 151
4.3 The model's interactions 162
4.4 The balance of power 169
4.5 The original question 172
4.6 The main direction 177

4.7 So what? Implications of the model 179
4.7.1 Person/people response 181
4.7.2 Consumer response 182
4.7.3 Worker response 184
4.7.4 Voter response 187
4.7.5 Business response 192
4.7.6 Education response 195
4.7.7 Society response 196
4.7.8 Government response 197

5 Conclusion – Technology as Artefact and Artefacts' Effect on Humanity 201

Notes 204

Bibliography 209

Index 211

Tables

2.1 Academic literature on the philosophy of technology 9
2.2 A critique of philosophies of technology 18
3.1 A typology of rationality in science 78
3.2 Fuel cell technologies 99
3.3 Fuel cell technology dependencies 100
3.4 A typology of medical technologies 102
3.5 An enlightenment time line 135
3.6 Productivity and social structures: how technology drives productivity which drives the human condition 142
4.1 Leading companies in global R&D 153
4.2 Dependencies in the model 165
4.3 Impact ratings in the model 171
4.4 Phenomena and technology determinism 174

Figures

3.1	The human symbiosis – nature mediated through science and technology via productivity	43
3.2	An illustrative complex network typology of technology	86
3.3	Declining infant mortality, 1964–2004	103
3.4	Male life expectancy at birth, 1964–2004	103
3.5	Female life expectancy at birth, 1964–2004	103
4.1	Distribution R&D expenditure by country	151
4.2	Distribution of G1000 R&D expenditure	152
4.3	R&D expenditure by sector and country	152
4.4	The technology value chain	156
4.5	Technology market interactions	159
4.6	Technology management business process	160
4.7	Technology business audit process	161
4.8	Dependencies in the model	164
4.9	The main causal direction	178

Acknowledgements

Thanks are due to GMDN Agency for permission to use Table 3.4 'A Typology of Medical Technologies' © Copyright and database rights: GMDN Agency Ltd 2005–2012. All rights reserved.

I also thank the Department for Business, Innovation and Skills for their permission to reproduce:

Table 4.1 'Leading Companies in Global R&D'

Figure 4.1 'Distribution R&D Expenditure by Country'

Figure 4.2 'Distribution of G100 R&D Expenditure'

Figure 4.3 'R&D Expenditure by Sector and Country' which are all taken from the UK R&D Scoreboard published by the Department for Business, Innovation and Skills, and are subject to Crown copyright status.

About the Author

Geoff Crocker holds a BA in Economics from Durham University and an MA in the Philosophy and History of Science from Bristol University.

Following an initial career with Rolls Royce UK, he worked extensively internationally, advising multinational industry clients in technology market strategies, including IBM, Yamaha, ABB, as well as a wide range of SMEs.

Over the past 20 years he has focused on the rapid development of the Russian industrial economy, working to develop and implement corporate strategies for major clients in many sectors of the economy.

This book combines academic content in the philosophy of technology with practical methodologies for business management of technology strategy. It is thus conceptual and practical, academic and managerial.

Geoff's wider work in technology market strategy is set out at www.tms.eu.com.

1
Introduction

Technology has immense impact on humanity. It certainly determines what we can do. Coupled with our decisions, our choices from the ever-widening possibilities it offers us, it therefore jointly determines what we do – how we live our lives. But more than this, it redefines who we are. We are no longer naked humans, but techno-humans. We don't have to wait for some futuristic science fiction scenario where we are all cyborgs. This fiction is already reality. We are already cyborgs. The only way we could survive without technology would be in a well-provided garden in a warm benign climate. It is interesting that the Bible opens with such a story. In this myth, humans lived naked in the Garden of Eden, and needed neither work nor technology. But as soon as they ate the fruit of the tree of know-how, they needed clothing, and were expelled from the garden. Reversing the logic of the story, as soon as they found themselves in a more hostile context, with less free-hanging fruit, then they needed technology. They had to work this technology to provide clothing, food and shelter. In his classic critique of 1930s Soviet life 'The Master and Margarita', Mikhail Bulgakov has the mythical Professor Woland magically clothe the credulous population of Moscow with luxury fur coats, only to remove them just as magically when they are later out on the streets, and so left cold and embarrassed in public in their underwear. Bulgakov's myth can also be interpreted for technology. It clothes us and makes us feel like lords, exuding pride and confidence; but we are entirely dependent on its magic. Its removal leaves us helpless.

Compared to other animals, humans have neither the stamina nor the strength to survive naked in nature. Without technology, humans would either not survive at all, or would be a very small species restricted to very benign parts of the earth, and only then if they could ward off powerful predators. But technology is an evolution which compensates for our limited human physical powers by harnessing our brain power instead. With it we have mastered much of nature. Technology is responsible for this huge shift of scenario. Its process is, as Vaclav Smil points out, more of a 'saltation' leap than a Darwinian mutation.

Technology therefore determines the human experience and the nature of humanity more than any god. Humanity could not exist in its present numbers and condition without technology. Basic needs of food, clothing and shelter for the current global human population of seven billion people could not be produced without the massive productivity enabled by technology. Even the simplest human joy is mediated by technology. The walk along the river bank is invaded by footwear technology and often by waterproof breathable clothing technology. The sail across the river uses computerised sail-shaping derived from advanced aerofoil technology, neoprene wetsuit technology, and fibreglass hull technology. The playing of the piano relies on vacuum-casting technology of precision plates. Apart from unaccompanied singing, music making depends on the specific technology embodied in musical instruments. Whole musical cultures derive from particular instrument technologies, from classical music depending on harpsichords and violins, to pop music born of the electric guitar and amplifier. These technology-driven musical cultures then foster specific associated social cultures. No aspect of the human life experience is technology free.

Without contraceptive technology many more human beings would have been born. Many more are being born with fertility technology. Without advanced medical technology, many more would have died. Technology has therefore fundamentally altered the profile of the human population and its life expectancy. It has also drastically altered humanity's quality of life. By taming and harnessing the elements, technology has shifted humanity from being subject to the cosmos and powerless before it, to being more or less its master. The shift is not total but it is substantial. Protection against wind, rain and sun is available, although earthquakes and tsunamis remain overwhelming. Massive exploitation of natural resources has advanced, challenged by more recent concerns that humanity has an equivalent responsibility as guardian of the environment, requiring the deployment of environmental technologies. Health has vastly improved. Diseases like smallpox have been totally eradicated, and there is technology in place to control other potential epidemics. At the same time, stress-related diseases have increased and the AIDS virus remains elusive, as does the common cold. Nevertheless, health outcomes are undeniably improved. Vastly fewer women die in childbirth and infants in childhood. Pain, where it occurs, is better controlled and mitigated, whether in dentistry, surgery, or in terminal cancer suffering. Some types of cancer are being overcome. Cancer is no longer the immediate death sentence it was only a few years ago. Gout is controlled and no longer leads to amputation. Antibiotics have greatly reduced bacterial infections, although recent resistant 'superbugs' are posing a renewed threat.

In the developed world, lifestyle has changed immensely. I remember growing up in a world without a refrigerator, telephone, television or car, and without central heating in the home. Life was still perfectly agreeable. Food was kept as cool as possible on a marble slab installed in the larder. Contact with other people was by personal visit and writing letters: less frequent and less intense. It was in my adolescent years when the first refrigerator arrived at our home, and a few years later when the first black-and-white television arrived, leading to a family Saturday evening ritual of sharing a bar of chocolate around a favourite television programme, a simple pleasure which today would seem very naive. We never got a car. A colour TV only arrived after I had left home for university. And the lack of central heating led to my being familiar with a steam-filled bathroom when a bath was taken. This caused me surprise later in life, when for the first time I took a bath in a heated bathroom. It took me a few minutes to work out why there was no steam condensing in the bathroom, allowing me to see unusually clearly across the room. Even in these micro-anecdotal personal examples, we can see how technology not only changes human practice but also human interactions, rituals, and patterns of life.

In economics, technology has hugely altered the production function, so that the number of person hours required to produce anything, and therefore to achieve any given standard of living, has fallen immensely. As a consequence, the production of a growing population has soared, leading to the rampant consumer age. It has changed economic and social structures. Feudalism gave way to democracy, modernity to post-modernity. It has thus also released time for people to develop arts and other interests beyond the requirements of mere survival. So life itself – the length of life, its quality and lifestyle – are all heavily determined by technology. This deserves the attention of philosophy.

We generally agree that technology has the fundamental and substantial roles in human life as outlined above, but we have very little understanding of how technology works in its interaction with humanity. In many cases we don't know how technology itself works. There is the well-worn joke in which the person who is the butt of the joke declares the Thermos flask to be the best invention of the twentieth century because it keeps some things hot but other things cold, whilst the user is bemused as to how the flask knows the difference! We are constantly using devices and systems technologies whose essential technique we don't understand. Car manufacturers badge their cars as having 16 valves, but very many, maybe most, people do not know what a valve does in a car engine and whether 16 of them is a good or bad thing. Here is an example of how cool post-modernity embraces the image, without bothering itself with the content. We really have become alienated from the technology which serves but at the same time fashions us. As a very minimum, we need a widely disseminated understanding of the technology process, and an awareness of, or at least a debate about, some fundamental issues concerning technology.

How much is technology a freely given endowment in that the natural processes it harnesses pre-exist, and how much is it a human creation, since humans have to isolate the natural processes used by any technology and re-configure them into the technology? So to what extent is technology discovered, and to what extent is it invented by humanity?

Secondly, has the process of technology developed 'a head of its own'? Even if we first created technology, is it now independent of us, some monster which now controls us, sometimes benignly, but at other times in a hostile threatening way? Can humanity now exert control over the technology process? Or is it a juggernaut that cannot be stopped? Does moral choice apply in the technology process, or is technology both all-powerful and totally amoral?

And does technology challenge the common materialist view that the universe is only physical? It isolates scientific processes from nature, and re-arranges them into reconfigured technologies which are then implemented in the physical world, but the core knowledge of this technology is metaphysical. The configured world therefore consists not only of matter, but of matter plus know-how.

1.1 The structure of a story, theory, philosophy and management of technology

Having set out these leading issues, the story line of technology worked out in this book unfolds the following structure

- **A comprehensive review of the academic literature on the philosophy of technology** is set out, with a summary of its strengths and weaknesses, and what relevance, meaning and significance it offers to our theme.
- **A network systems model of technology in symbiosis with humanity and nature** through the artefacts of science, the economy, productivity, and society is set out.
- The **assumptions** stated for the model are then examined in greater detail as are the **entities** of nature, humanity, science, technology, economy, productivity and society central to the model. Some creative interpretations are considered for further discussion and incorporation into the model.
- **Technology** is then set into a **typology** which is exemplified with a selected range of technologies. This section is largely a **narrative** of technology, partly to establish the descriptive story which forms the set of observed phenomena for the theory developed in the model, partly to generate issues from this narrative observation, and partly as of interest per se.
- **Business management models of technology** are then advanced, including a methodology for the development, evaluation and implementation of technology market strategies, and proposals for reporting of business value according to technology portfolio holdings.
- **Social models of 'innovation policy'** are then discussed.
- Finally, the model is synthesised through various iterations to generate conclusions which are stated as a working hypothesis for a **systems network managerial philosophy of technology.**

We now turn to a review of the academic literature on the philosophy of technology to see how far the questions set out above are adequately addressed and maybe even resolved.

2
The Academic Literature

Given the immense core effect technology has on humanity, it is surprising that philosophy pays little attention to it. The academic literature comprises only a small corpus, the bibliography to this book comprising virtually its entirety. Even flourishing philosophy of science schools rarely mention the philosophy of technology, despite its primary role in mediating science to humanity, and its evident importance to the human condition and experience. What mention technology does get is often mega and negative. Martin Heidegger and others' fear of nuclear technology – with its capability to destroy the earth several times over, or Rachel Carson's concern with agricultural fertiliser and pesticide technology set out in her 'Silent Spring', are well known examples and deserve debate. However, the more regular, less exotic but more complex and interactive effect of technology on humanity has been relatively ignored. Because we have not so far suffered a global nuclear holocaust, or the total destruction of our food crops from overdose of fertiliser and pesticide, we have become complacent towards the power and effect of technology. Its effect may be micro rather than mega-macro. It may be hidden deep below the surface of the products we consume. It may be benign, delivering high standards of living through hugely increased productivity. But it is extensive and pervasive. We need to understand its processes, rather than let it lead us wherever it will. We need to know whether and how we can lead and manage technology.

There is a divide, and almost an antipathy, between classical philosophy and technology, rather than the convergence of interest which is needed and justified. This is expressed in some frustration by the authors of an embryonic academic philosophy of technology. Val Dusek opens his introduction to the subject by noting that 'Only sporadically were there major philosophers who had much to say about technology, such as Bacon who in 1627 in his "New Alantis" 'imagined wise men in a house of philosophy (Solomon's house) applying philosophy to the mechanical arts', and Marx in the mid-nineteenth century.[1] Most of the 'great philosophers' of this period, he says, 'although they had a great deal to say about science, said little about technology.' He becomes even more critical of the nature of what philosophy of technology does exist, referring to 'a highly convoluted and obscure European literature' which is 'grand and ambitious, but often obscure and obtuse' saying of Heidegger, Arendt, Marcuse and Ellul whose work we review later, that they are 'all notorious for the difficulty and obscurity of their prose'.[2]

Olsen and Selinger, in their introduction to a series of interviews with 24 inter-disciplinary writers contributing to the philosophy of technology, note that 'only a few professional philosophers actually specialise in the philosophy of technology properly' and that 'it is difficult to find an institutional core canon of texts and figures that rigidly defines membership or methodology'.[3] Trevor Pinch and Wiebe Bijker in their development of social constructivism as a philosophy of technology, write 'Indeed the literature on the philosophy of technology is rather disappointing'.[4] Albert Borgmann in his interview laments 'the danger ... that the high analytical style of philosophy will continue to sap the philosophical concern with the concreteness of life and the reform of society. Philosophy of technology needs a conversation not just with the philosophical elite, but as much with social scientists, fiction writers and journalists'.[5] Mario Bunge attacks Heidegger's 'pseudophilosophy',[6] accusing it of failure to be 'clear and subject to rational debate'. There are, he says, 'uncounted philosophy teachers for every original philosopher'.[7] The philosophy of technology as a relatively new field needs original philosophers rather than teachers of others' philosophy. The lack of a strong cohort of original thinkers may itself account for the continuing low profile and underdevelopment of a philosophy of technology.

We have established several leading issues for a philosophy of technology against which the academic literature should be searched and evaluated. First, technology must be defined partly as a freely given endowment to humanity, in that the natural processes pre-exist, and partly as the product of humanity's effort, first to isolate the natural processes, then to harness them in a different configuration of technology processes.

Second, there is a clear symbiosis between humanity and technology. The question is how this symbiosis works. On the one hand, humanity researches science, to harness the way the natural world works to engineer technology, whilst on the other hand, technology impacts humanity in its turn. The metaphysical question is whether in this symbiosis technology has itself become a reified artefact, an independent 'thing' – a power existing separately from humanity, although created by humanity. Does technology now have independent status, determining the fundamental nature and essence of humanity, the human condition and human life?

Third, technology can well be defined as 'the cognitive reconfiguration of natural resources and processes'. But 'reconfiguration' and 'process' are metaphysical concepts. This necessarily renders the nature of humanity as not just physical, but a physical + metaphysical artefact. This conclusion harmonises and agrees with the necessary metaphysical definition of more general human ideas and emotions. Ideas and emotions can be multiplied instantly in many physical hosts, the human brain or a computer disk, without any increase in mass. They are entirely dependent on their physical host for their existence, but nevertheless once extant, they have independent status. The current dominant philosophy of physicalism without 'qualia', analysed so exhaustively and widely popularised in Frank Jackson's book 'There's Something about Mary', will not suffice. It is an insufficient ontology of techno-humanity. Again, the absence or paucity of any philosophy of technology in university philosophy departments, to address these issues, is startling.

The following sections offer an analytical review of the major academic literature in the philosophy of technology. An analysis showing the major contributors at each point in the structural development of the literature is set out in the facing page.

After setting out these contributions descriptively, we then proceed with a critique and an alternative reformulation.

2.1 Definitions of technology

2.1.1 Rationality

Andrew Feenberg is a leading contemporary philosopher of technology. His cycle of major publications runs through his 'Critical Theory

Table 2.1 Academic literature on the philosophy of technology

	Contributor
Definitions of technology	
Rationalisation	Andrew Feenberg
Human enhancement	Philip Brey, Nick Bostrom
Cyborg ontology	Donna Haraway
Instrumentalism + intentionality vs substantive view of technology	Mario Bunge, Keekok Lee, David Kaplan
Philosophical analysis of technology	
Determinism	Karl Marx, Robert Heilbroner
Autonomy	Jacques Ellul, Alfred Borgmann
Social constructivism	Trevor Pinch and Wiebe Bijker critic Langdon Winner
Technocracy	Andrew Feenberg, C P Snow
Dystopia	Martin Heidegger, Herbert Marcuse Hans Jonas
New conceptualisations of technology	
Phenomenology	Don Ihde
Moderation of technology	
Saving grace	Martin Heidegger
Catholic faith	Jacques Ellul
Focal things and practices	Alfred Borgmann
Democratisation	Jorgen Habermas, Andrew Feenberg
Correctly conceptualising technology	Heidegger suggests but fails to define. Therefore a definition is offered in a 'Systems Network Philosophy of Technology'

of Technology' 1991, 'Alternative Technology' 1995, and 'Questioning Technology' 1999 to which he has recently added a collection of his papers in 'Between Reason and Experience' in the MIT 'Inside Technology Series' 2010. Graeme Kirkpatrick has written a major study covering Feenberg's work 'Technical Politics: Critical Theory and Technology Design'.

In his review of Feenberg's work, Philip Brey shows that Feenberg defines technology as 'the sum of rational means employed in a society',[8] a definition which is critiqued below. For Feenberg, technology is coincidental with modernity, which Feenberg defines as 'the project of building a rational society', following Max Weber who saw western society as the rise of rationality, the incarnation of the Enlightenment, the Age of Reason. Thus 'rational systems such as technology, science, markets and law play a privileged role in modern societies'. Feenberg however believes that there are no universal principles of rationalisation, but that the specific rationalisations which emerge are 'contingent': that is, they depend on what gave rise to them, are relative not absolute, could have been otherwise, or are not necessarily so. Alternative rationalisations are therefore possible; an assumption which allows Feenberg to develop a critique of current modernity and propose an alternative modernity, principally one where technology is politically contingent and can then be made subject to a process of democratisation.

There is some confusion of terminology here. In one sense rationalisations are generic and necessarily so. For example, evaporation causes cooling through high energy molecules becoming vapour, leaving lower energy molecules in the remaining liquid and therefore the liquid cooler. This is a generic rationality – a connected trail of cause and effect. From it we build refrigerators, which always work because the 'rationality' that evaporation necessarily causes cooling is generic and unavoidable. Only if we call refrigeration itself the rationality, that is, shift the definition of rationality to the reconfiguration of rational cause and effect processes, is Feenberg justified in saying that rationality is contingent, and that we have potential choice, for example of a refrigerated kitchen world or not. We do not have a choice as to whether evaporation causes cooling, but we do have a choice as to whether to deploy this technology in refrigeration, and we can apply rational criteria to this choice.

There are further potential objections to the singularity or primacy of rationality as a definition of technology. One is brought out by Val Dusek who, writing further on Max Weber, says that 'The goals or values about which the (instrumental) means are rationally structured, are based on irrational decision. There can be no genuine reasoning about values. Weber agrees with the existentialists and sees choice of values as an arbitrary, irrational decision.'[9] Rationality can therefore apply to the means of technology, but not to its objectives or its values. Convincing as this sounds, this argument is not necessarily true. We are economic as well as technological beings. Our economic choices do, at least sometimes, have rational foundations. So whilst we understand the rational cause and effect principles and processes behind refrigeration, we also have rational grounds for deciding whether or not to deploy refrigeration. Conservation of food along the food chain is the major rational basis for refrigeration. In countries like India, with low levels of distributed refrigeration, some 30 per cent of food production is lost before it reaches the consumer's home due to inadequate refrigeration along the value chain. We understand (almost) how nuclear technology is applied to heat water, drive steam turbines and generate electricity, but we also consider rational factors in deciding whether to deploy nuclear technology, including its high cost of spent fuel processing and plant decommissioning, set against its emission-free status compared to fossil fuel power-generation technologies. There are rational factors in choosing between private car and public transportation technologies. What is not rational is the weight we attach to any of the rationally identified outcomes. We can only compare the faster door-to-door time and the private ambience of car travel with its higher carbon footprint by attaching subjective weights to these outcomes. The weights are subjective but their application is still rational. If objectives and values are allowed an independent ontology, then they have to be defined outside technology and brought to bear in symbiosis with it.

The other objection is that humanity is not just rational but also holistic, with valid emotional and spiritual components. As the Romanticism which followed the Enlightenment, and the post-modernity which followed modernity have demonstrated, humanity incorporates a 'feel factor', has feelings as well as thoughts, emotion as well as rationality. In the humanity-technology symbiosis, technology must therefore be synthesised with the holistic nature of humanity. Much post-modern thought seeks to debunk the hegemony of rationality entirely, which is an initiative destined to failure given the apparent objectivity of deductive logic (more on this later). Nevertheless, rationality is synthesised with emotion in holistic humanity, and this feeds into the objectives and values set for technology, since technology only exists in symbiosis with humanity just as humanity only exists in symbiosis with technology.

So technology is inherently and intrinsically rational and has strong elements of rationality in its application, but is ultimately subject to a subjective feel factor or value set. Technology is only rational because the natural processes it harnesses happen to be rational – that is because deductive logic is objective. If nature were not rational, then technology could not be defined as rational. Some more recent post-Newtonian views of nature indeed question its rationality. The Copenhagen interpretation of quantum mechanics implies that particle positioning is stochastic rather than deterministic. The cause-effect connection is not one to one but one to many. Defining technology as the cognitive reconfiguration of natural materials and processes, survives the inclusion of non-rational aspects of nature in the way the definition of technology as rationality does not.

2.1.2 Human enhancement

The anthropological definition of technology is that it adds to human power and capability. This is the definition of technology as tool. The techno-human is simply an extended human. Stephen Kline defines technology as 'extensions of human capacities by socio-technical systems'.[10] Historically he sees this process as having taken a quantum leap forward since 1840, which closely aligns with Martin Heidegger's differentiation between traditional humanised hand technology and modern systems technologies, and with Keewok Lee's view that up to 1850, technological development was largely opportunistic happenstance, but then became determined and science-led.[11] Kline is emphatic about the crucial importance of technology in enabling and thereby defining humanity, saying 'Without socio-technical systems, we humans might not exist as a species, and if we did, we would be relatively powerless, few in number and of little import on the planet – we have in a large measure become lords of the planet.'[12]

Nick Bostrom, in an article on the future of humanity, shows how technology has accelerated human/world development. He points out that the period from eukaryotic life to prokaryotic life spanned 1.8 billion years, the era of multi-cellular organisms 1.4 billion years, whilst homo sapiens has existed for only 100,000 years and has grown from a population of 5 million at the time of the agricultural revolution to 1 billion in 1835 and 6.6 billion in 2007.[13] Will technology at some point enable a 1 trillion human population, with a life expectancy of 500 years, a vast increase in cognitive capacity, and total environment control?

Arnold Gehlen analyses this anthropological definition of technology in his article 'A Philosophical Anthropological Perspective on Technology.'[14] He refers back to Ernst Kapp's 1877 'Philosophie der Technik' where Kapp makes the connection between 'man's (sic) organic shortcomings and his inventive intelligence'. The necessity for technology derives from man's organic deficiencies because 'man in any natural uncultivated environment is not able to survive because of a lack of specialised organs and instincts – he has to create the conditions for his physical survival by intelligently altering existing environmental conditions'. Kapp speaks of human organic relief, substitution, replacement, strengthening and improvement from technology. This definition could well incorporate the modern potential for enhancement through genetic modification. Gehlen quotes Freyer as saying that this technology creates a kind of abstract ability.[15] a forerunner to later interpretations of technology as a fully reified artefact, responding in symbiosis with humanity – not necessarily benignly but potentially threateningly.

Technology, which began as an extension of human power, has subsequently objectified first human physical power, and then human intellectual power. Man is now 'a feedback system which reacts to its own products'.[16] Gehlen asks what are 'the consequences of this externalisation of our functions for man's self understanding?' Technology has become 'a dubious expansion of consciousness'. He quotes Heisenberg in a 1926 lecture saying 'technology in fact no longer appears as the product of a conscious human effort to enlarge material power, but rather like a biological development of mankind in which the innate structures of the human mechanism are transformed in an ever-increasing measure to the environment of man in a biological process which is no longer subject to human control'.[17]

Whilst human enhancement might suffice as an understanding and definition of technology for traditional hand-tool technologies, as in Heidegger's example of the hammer, it will not suffice to represent modern highly complex technology systems which generate wider issues of the autonomy and determinism and/or social constructivism of technology. It is a micro definition which is not relevant to macro technology. Nevertheless, as Gehlen shows, it still opens the way to the conceptualisation of technology as an independent threat to humanity.

Is this definition of technology as an extension of human power adequate? As a definition it is true but incomplete. First we have to ask what it is that is extending human power. Without answering this question we have hardly defined technology at all. It is the application, the reconfiguration, of natural materials and processes with which technology has extended human power. This is therefore a better definition of technology, that it is the cognitive reconfiguration of natural materials and processes.

Technology *is* an extension of human power, but is not *only* an extension of human power. It is also an investment of human intelligence, with reconfigured materials and processes from nature, into a separate entity. The hammer is simply an extension of human power. The first hammers were simply hand-held lumps of stone. But even their application required know-how, or technology, of how to use the stone, for example to drive a wooden stick into the ground. Modern hammers, it is true, are little more than the metal equivalent on a wooden handle, and therefore an extension of human power. The hammer remains totally subject to the human actor. It is essentially an extension of the human arm and human arm-power. When we look at other technologies, such as telephony or power generation, this analogy becomes less clear. Physical power is not the sole measure of capability. Complex systems embody human intelligence, and extend that intelligence in various dimensions, such as calculation speed. But in other dimensions of human intelligence, technology limits rather than extends human power. This is the case with AI, artificial intelligence, a technology which Dreyfus claims has delivered a reduction of human power. What is important is the potential independence of the technology from its human creator. The hammer is totally dependent on humans for its action. Complex technology systems are not; they are capable of independent operation. They are thus not simply an extension of human power. They may even be a reduction of human power. But, once implemented and installed, they are independent of real-time human power at the time of their operation. It is this characteristic which goes beyond the definition of technology as tool and requires a different definition. Technology may have independence of operation, not only independence in its development trajectory.

2.1.3 Cyborg ontology

Stephen Kline's comment about the extreme vulnerability of humanity without technology is brought to its ultimate conclusion in Donna

Haraway's work. Haraway defines technologised humans as 'cyborgs', writing 'By the late twentieth century, our time, a mythic time, we are all chimeras, theorised and fabricated hybrids of machine and organism; in short we are cyborgs. The cyborg is our ontology'.[18] This may sound like a term borrowed from science fiction but, in reality, since we admit that humanity could not exist without technology, then an ontology of standalone humanity is insufficient and redundant. Only an ontology and philosophy of techno-humanity, which Haraway labels 'cyborg', works. Understanding the nature of the symbiosis of humanity and technology then becomes a sine qua non of all philosophical endeavour – one towards which this book seeks to contribute a framework.

We do not need to suffer physical change to be cyborgs. Some technologies may indeed invade or attach to the human body. This is true of medical technologies, whether drugs to control blood pressure, gout or thrombosis, or electromechanical devices such as heart pacemakers. Use of these renders us techno-humans, as does the wearing of glasses for reading, or infra-red glasses for night sight. But we are all clothed most of the time, whether for climate reasons, or to conform to convention. We quickly get used to how people dress, even though history shows that styles have changed dramatically. We therefore consider a clothed human normal, which is statistically true, but in fact a striking truth, since no other species clothes itself, either of necessity, or of propriety, or of vanity. So how do we define a human being wearing clothes, using reading glasses and maybe a hearing aid, taking drugs to control body functions, fitted with a pacemaker, and eating processed foods? Surely this is a cyborg?

In seeking definitions however, the cyborg terminology does not define technology. It redefines what technology has rendered humanity. It is a word meaning 'techno-human'. Technology could be defined as 'that which adds to a human to make a cyborg', that is, mathematically, cyborg minus human must therefore be technology. The preferred definition of technology as the cognitive reconfiguration of natural materials and processes retains its force.

2.1.4 Instrumentalism

The important aspect of the instrumental definition of technology is its focus on human intentionality. This is different to and goes beyond the above view of technology as tool since the tool can be intention free. Technology, according to the instrumental view, is a means to an end. From this perspective, technology is value free, and all values rest in the human intentionality to which technology is harnessed to various ends. Thus Benjamin Hale writes that 'technological artefacts already embody moral consideration in their creation'[19] and claims that 'moral status is best understood as a question for the agent'.[20] On the other hand, opposing this, Peter-Paul Verbeek suggests that, whilst modernity's dualism of subject and object implies the separation of human agency from technology, contemporary reality is 'a web of relations between human and non-human agencies'[21] within which 'technology appears to be able to act in the world' thus departing from 'a human monopoly on agency'.[22] In this framework technologies can be coercive, or at least persuasive and seductive. The instrumental definition, however, sees technology as subordinate to humanity, which differs entirely from substantive views of technology such as the deterministic and social constructivist analyses reviewed below.

Mario Bunge defines technology as 'the control or transformation of reality, whether natural or social'.[23] Keekok Lee echoes traces of Heidegger in defining technology as 'the instrumentalisation of nature, an instrumental attitude towards nature'[24] although he sees it as anthropocentric, which is distinctly non-Heideggerian when considering complex systems technology. According to Lee, artefacts in general, and technology in particular, are 'the embodiment of human intentionality' and technology is the material embodiment of the same human intentionality. David Kaplan writes that 'technology incorporates two conceptualisations of the world: one physical, one intentional'.[25] The instrumental view of technology focussed on intentionality necessarily incorporates causality, since technology is the means to the end specified by intentionality. It is therefore a combination of the definitions of technology as rationality, human enhancement and human intention, but does not define technology as an independent reified artefact. This view is naive, and is highly contested by analytical philosophical views which define technology independently of humanity and therefore as a potential threat to humanity.

Technologies can emerge as a result of human intention, but they can also arise independently of human intention. Human endeavour may seek solutions to an end objective, for example, telephony to enable speech at a distance. However, once this technology is developed with its switches and wires, it becomes evident that the same technology can support data and image transfer. Applications are therefore enabled beyond the original human intention in developing the technology. Such unexpected outcomes can even contradict the original human intention, as was the case in the development of dynamite for which Nobel had only constructive intentions, and so lamented its unintended application in warfare and killing. If technology can turn out to surprise and even contradict the human intention, then it cannot be defined in this simple instrumental way as subject to human intentionality and agency.

The development of technology can thus breach human intentionality, but is this also true of its application? It is more arguable that the application of technology is subject to human intentionality. Nuclear weapons have been used twice to devastating effect. The post-war consensus against their further use has held, although current timescales are trivial. For less obvious and less highly profiled technologies, their application is implemented without widespread knowledge. In some cases, there may seem little choice in adopting a technology. If a new technology enables substantial reduction in employment, competitive market pressures may give producers of a good or service little choice in its adoption if they want to survive. It may not at all be their intention to reduce employment, but the technology has the power to insist on its adoption. Technologies available to turn coal into oil have large economic potential when oil prices are high, but they require huge supplies of water and emit enormous quantities of CO_2. Humans may not intend to emit any more CO_2 but may decide to implement coal to liquid technologies for political and economic gain. The detailed nature of the technology then means that humanity has ended up doing something it did not intend, in this case, more CO_2 emissions. In other cases I may not intend to travel by car, but if car technology has found so many other adopters that public transport is rendered uneconomic and is withdrawn, then the technology nexus has forced me to travel in a way I did not intend.

Even in application therefore, technology cannot easily be defined according to the instrumental definition which leaves it subject to human intentionality.

We can summarise our presentation and discussion of the above four tentative definitions of technology as

Table 2.2 A critique of philosophies of technology

Definition of technology	Critique
Rationality 'the sum of rational means employed in society'	• Need to distinguish intrinsic rationality of the technology from the rationality of its application • Application of technology can be rational too • Technology is rational only because nature is rational • So rationality is a definition of nature, not of technology • Humanity is holistic, not only rational • Humans have subjective, even irrational, values and objectives • So since humanity engineers technology, technology may not be totally rational
Human enhancement Technology as tool	• OK for hammer but not for complex system • Need to define **what** extends human power that is, not a definition at all • Technology can reduce human power, example, AI according to Dreyfus • Once operationally independent of human power this definition fails
Cyborg Humans are techno-human	• A definition of humanity, not of technology • 'Cyborg minus human= technology' doesn't work
Instrumentalism Technology subject to human intentionality Technology is value-free Human agency is determinative	• Technology **developments** can surprise and even contradict human intentionality • **Applications** of technology can also contradict elements of human intentionality • Technology adoption is often unnoticed, that is, hardly intentional • The technology process is more substantive than instrumentalism allows

For these reasons the preferred definition of technology is

'The cognitive reconfiguration of natural materials and processes'.

This definition will be worked through into a network systems philosophy of technology in a later chapter.

2.2 Analytics of technology

The above definitions of technology see technology as instrumental, dependent on human intention and/or human understanding and application of rationality. In these definitions, technology is subject and subservient to humanity. It cannot act independently. It is therefore never a threat to humanity in general, although one subset of humanity could clearly adopt some technology as a threat to another, as has frequently been the case in human warfare.

More substantive definitions of technology are derived from philosophical analysis of the interaction between technology and humanity, where technology gains the quasi-independent status of an artefact which can impact on humanity.

2.2.1 Determinism

The leading question here is whether technology determines humanity and/or human society, and if so in what way, and how this determination works in practice.

As far back as 1620, Francis Bacon already viewed technology as having impact. He writes in Novum Organum 'Again we should note the force, effect, and consequences of inventions, which are nowhere more conspicuous than in those three which were unknown to the ancients; namely printing, gunpowder and the compass. For these three have changed the appearance and state of the whole world; first in literature, then in warfare, and lastly in navigation; and innumerable changes have been thence derived, so that no empire, sect or star, appears to have exercised a greater power and influence on human affairs than these mechanical discoveries'.[26]

Later a similar view was taken of optics technology, specifically the microscope and telescope. These allowed humanity to see worlds beyond their naked vision, and to build data on phenomena at cosmic and microcosmic levels which then allowed scientific theorising and technological intervention. There is no doubt that the microscope and telescope have been fundamental in developing a wide range of technologies, from engineering to medical technologies, and have therefore hugely impacted human life.

In these examples, technology acted at the level of device, and therefore in one sense seemed entirely subject to human control. But the innocence of the device does not eliminate the potential independent power of the technology process of which the device is a mere component outcome. Would it have been possible for humanity to consider the microscope, project the totality of all its future possible applications, and the technologies to result from those applications, and make a decision whether to adopt the microscope or not? Probably not. So the optical technology incorporated into the microscope is determinative. It would be virtually impossible for humanity to ban the microscope, both because of the data needed to make such a decision (i.e., the discounted value of all future microscope-led outcomes), and the impossibility of policing such a ban. Technology has an almost intelligent way of finding other human adopters if some part of human society rejects or ignores it. It is rather like water: it will always find a course to run. We are already seeing, even at the level of device, ways in which technology is potentially an unstoppable process. If indeed we cannot control technology, or at least cannot easily control the technology process, then we should at least be aware of this, and take it into account when we think about technology and interact with it.

Karl Marx's famous comment in his 1847 'The Poverty of Philosophy', that 'The hand-mill gives you society with the feudal lord; the steam-mill society with the industrial capitalist'[27] underlines Marx's view that technology is the servant of capital and drives the economy from feudal into capitalist structures. This takes the concept of technology determinism even further. Not only are we likely to adopt technological devices such as the microscope and mutely accept all their unforeseeable outcomes, but if Marx is right, our whole social structures are not chosen by us, but determined by technology. This again deserves careful consideration and debate. Are feudal societies a necessary concomitant of an agricultural economy? Could we imagine agricultural capitalism? Presumably we could, since the agricultural sector exists within multi-sectoral capitalist economies. So it doesn't seem that an agro-economy demands feudalism per se. On the other hand, does industrial technology require capitalism? An industrial economy certainly does require some different social organisation to that of an agricultural economy. Its method of production requires workers to gather into factories. These factories become very large people employers and so drive the development of cities which also become the consumers of the industrial economy's product. Capitalism also requires capital, and this requirement was on such a large scale that it could no longer be provided by a single individual feudal lord. The joint stock company system, designed to meet this need for large-scale capital investment, meant that ownership of the place of production, and therefore of the place of work, was no longer solely in the hands of the local lord as it had been in the agricultural economy. Social structures were therefore depersonalised. In this sense feudalism transformed into capitalism and the partitioned lord/serf social structure transformed into the upper class/working class divide of capitalist society. Classes replaced individuals, and this is the essential transformation from feudalism to capitalism.

However, a common cause was also at work. The same Enlightenment which led to scientific enquiry and technology development also enlightened social values. Campaigns against serfdom and slavery accompanied the outbreak of the rational pursuit of science and technology, as well as greater free expression in the arts. In 1791 Thomas Paine published his 'Rights of Man', arguing against the hereditary principle in government and in favour of democratic government. He was sentenced to death for this, but made his escape to France. In 1775 Thomas Spence had argued in his lecture also titled 'The Rights of Man' for a more communist rather than democratic government. This all demonstrates that the awakening of the Enlightenment had widespread effects in many areas of human life: changes in technology were accompanied by changes in social thinking and the philosophy of humanity, the nature of human rights etc. So technology may not have caused change in social structure, but more have shared a common cause with it and was therefore inseparably bound to it.

In support of this view we can see that capitalism does have many and different associated social structures. In the USA, capitalism is very entrepreneurial with less state involvement than in Europe, where more socialist elements in the concept of government combine with free-market enterprise. In Japan, capitalism is subject to the respect of social structures; for example, respect for the older person regardless of the rationality of their decision. Korean capitalism is expressed in more militaristic terms. Huge 'chaebol' business conglomerates dominate the capitalist landscape. In Russia, a specific form of capitalism has been created from the previous communist state industrial structure. The new Russian shareholders hold very concentrated power, compared to the widespread fragmented shareholding typical of western companies. Often, one or two individuals control large dominant companies, for example in telecommunications, steel or energy, and these industrial barons exert a more feudal and less democratic power. Feudalism can therefore co-exist with industrial technology, in this specific and unique Russian case, only because the need for large-scale capital investment beyond the resources of the feudal lord, which led to the joint stock company in Europe and the USA, was by-passed and undertaken by the Stalinist state, which industrialised the economy then later collapsed, leaving its pickings to latter-day feudal industrialists.

This all suggests that technology does not necessarily drive and determine social structures. It does demand new workplace configurations, and the upheaval of urbanisation which this causes does lead to new social structures, and gives opportunity to the development of new social philosophies. In the personalised setting of agricultural feudalism, disseminating revolutionary creeds could be difficult and tightly constrained. But in the depersonalised anonymous urban industrial setting, such new philosophies of democracy, socialism and communism could be generated in response to the conditions of life, as they were in Frederick Engels' 'The Condition of the Working Class in England in 1844'. They could also be more easily promulgated.

As has also been outlined above, technology does share a common cause of intellectual awakening with democratic development. Technology and social change therefore go together, although their co-existence is not necessarily the result of a direct causal link.

The seminal contemporary contribution to the examination of the hypothesis of technological determinism is Robert Heilbroner's celebrated, short and very readable essay 'Do Machines Make History?'[28] Heilbroner sets out examples of military technology determining outcomes in warfare, TV and Internet technology having political impact, and technology's effect on production affecting the socio-economic order. Technology therefore definitely has impact on humanity. He then asks the core question as to whether 'technology is the prime mover of social history?' and whether 'we can explain the laws of motion of technology itself?' He points out that there are no empirical studies testing the hypothesis of technological determinism, but he does take the view that there is an inescapable sequence in the development of technology.

Heilbroner believes that technology follows an inevitable development path. Each stage along this path is dependent on the former stage and cannot happen without it. The stages are necessary and fixed. The exact history of technology could not be otherwise. He shares this view of the path dependency of technology with Nathan Rosenberg's exposition in his 'Technology and the Wealth of Nations'. His justification for the view of an inevitable path development nature of technology is

i) the frequent simultaneity of technological discovery
ii) the absence of technological leaps
iii) the predictability of technology

He accepts that these observations do not conclusively prove his view of an inevitable necessary path-dependent evolution of technology, but he argues that they support the idea. He admits that there are no empirical studies against which these hypotheses can be tested, nor does he propose any. Twentieth-century history does give examples of simultaneous technology developments. For example, Russian scientists were working on penicillin in parallel with Alexander Fleming; US and German military technologists were both developing the nuclear bomb. With today's Internet networking of scientists and technologists, global collaboration on technology development is high. Multinational corporations are responsible for technology research and application globally. These both mean that there are fewer separate communities and societies to test Heilbroner's hypothesis of the simultaneity of technological discovery.

Heilbroner's definition of technological leaps is subjective. What is evolutionary, and what is a quantum leap? This makes his second point also difficult, if not impossible, to test empirically. His further claim of the predictability of technology is interesting but contentious. It is again a hypothesis which is difficult to test empirically, as the test would have to last several years, and the degree of conformity of technology's outcome to its earlier prediction would have to be defined, a definition which is in danger of tautology.

We may agree with Heilbroner that one technology development depends on a preceding one. The most frequent case of this is the previously quoted example of the microscope, which has undoubtedly enabled, but also therefore been essential to, many subsequent technologies. However we cannot be certain that because technology B required the previous establishment of technology A, that technology A therefore necessarily leads to technology B. The technology development path could have been otherwise. Technology A could have led to technology C or Z rather than technology B.

Heilbroner essentially agrees with this, but implicitly rather than explicitly. He does this by accepting that his path-dependent technology evolution is also constrained. The constraints he lists are

i) the accumulated stock of available knowledge
ii) material competence, what Heilbroner calls the 'necessary requirement of technological congruence'
iii) the size of the capital stock
iv) the specialisation of labour

To this list could be added

v) the background science which enables technology but therefore also acts as the constraint on its possibility

Any one technology can only be implemented in the context of a background of enabling conditions, including other technologies. General know-how has to be available. Specific engineering capability has to implement the technology; Heilbroner gives the example of steam-engine technology needing metal casting and welding expertise. Indeed, the lack of advanced skills in boiler manufacture led to numerous dangerous failures of early steam engines. The capital stock of machine tools and factory manufacturing capacity has to be sufficient to deploy the technology, especially when it is a product technology. And labour has to be able and willing to operate the technology, whether this is an electron microscope, an Excel spreadsheet, or a computer-controlled machine tool.

These are undetermined parameters. The result of their constraining the evolutionary development of technology is that this evolution is not unique or necessary, but contingent, depending on the state of these constraining parameters in any one society at any one point in time. Heilbroner is of course unable to subject this more complex model to empirical test. The concept of such a model, with technology as the dependent variable, tested against independent causal variables of previously implemented technologies and the five constraints set out above, is difficult to imagine, specify or develop. Heilbroner's hypothesis therefore remains interesting and heuristic, but speculative.

Heilbroner notes that technology does change the composition of the labour force and the hierarchical organisation of work, and concludes in favour of a 'soft determinism' theory of technology. He resists Marx's unilateral view of technology's impact on society by pointing out that technological progress is itself a social activity, responds to social direction, and must be compatible with existing social conditions. Different societies manage the technology outcome differently; Heilbroner quotes contemporary Kalahari tribal society, former tenth-century technocratic Arab societies, and Chinese society. Presumably, contemporary Kalahari tribal society could choose to gradually implement advanced global technology, but in so doing would no longer be Kalahari tribal society. The technology would require a different social structure, both in production and in its consumer use. So technology sets and the specific structures of specific societies are linked – they are matching pairs in technology/society equilibrium. But the exact pairing remains contingent; other equilibrium pairings are equally possible. In this sense, technology does not uniquely cause social structure, nor does social structure uniquely generate technology, but the two interact and have to find an equilibrium pairing. Such equilibria are not total or permanent, because shifts in the technology, and potentially in the social structure, drive a dynamic process in which the technology/society pairing rapidly evolves.

For Heilbroner, technology and society are indeed interactive. He suggests that the rise of market capitalism provided a context in which technology could flourish and permeate the human condition almost, apparently autonomously, but he sees this as peculiar to a period of high capitalism when 'the agencies for the control and guidance of technology are still rudimentary'. He clearly believes that social control of technology is possible and expects that it will be exercised, much as other writers who urge the democratisation of technology. This is a claim we will examine more thoroughly later.

Andrew Feenberg, for example, denies that technology is either determined or determining. For Feenberg technology is 'politically contingent', a view he needs for his proposal for the democratisation of technology examined below.

2.2.2 Autonomy

The ultimate determinism is one where, contrary to Heilbroner's view, technology is not interactive with society but is autonomous, dictating outcomes to humanity. Here we encounter the theory of technology as threat. The question is whether technology is 'objective', that is, whether it exists, operates and evolves independently of human action and intervention. Nature is objective in the sense that the planets pursue their trajectories independently of humanity. Science is only partially objective in that, whilst its discovered content is of objective processes within nature, the attempt to isolate and understand those processes and to form scientific theories about them, is a human decision and priority. Potential science, defined as the eventual totality of all processes in nature, is objective, but actual science, which is the human-determined subset of this, is not.

The philosophy of mathematics suggests that mathematics is objective, mainly via the argument of 'epistemic constraint' that, for example, it appears that there is an infinite number of prime numbers but it is impossible for us to know them all. Therefore, mathematics is discovered rather than invented. It has exogenous existence. This conclusion is not certain, since it can be argued that mathematics is mind-dependent, invented rather than discovered, and that it is compatible with this view that an infinite number of prime numbers have been invented without our being able to know them all, that is, that the epistemic constraint does not prove the objectivity of mathematics. Deductive logic also seems strangely to be objective, although how this can be so is a supreme puzzle. It is therefore theoretically a priori possible that technology, either as content or as process, is also objective.

Technology as content does have some objective qualities in that it entirely derives from reconfiguration of natural physical material and natural processes discovered by science and harnessed by technology. Technology as process may well also have some autonomous element but also an element which is contingent on human decision and action. Jacques Ellul analyses this question in his 'The Technological System'.[29] He quotes K Pomain in 'La Malaise de la Science' that 'all science is implicated in the technological consequences' and 'all science having become experimental depends on technology'. Science and technology are therefore in interactive symbiosis. Heidegger thought that technology led science in the symbiosis, but I will argue later that science necessarily has the role of primary mover, since I define science as both 'knowing that' and 'knowing how'. Ellul asks the pertinent question that if technology is thought to be autonomous, what is it that technology is claimed to be autonomous from? If from the economy, from industry, from political ideology, then he concludes that these relationships are bi-directional and interactive, but that technology can certainly initiate a shift in each relationship and hence has independent status. In its interaction with ethics, autonomous technology can 'render us amoral'. We will consider later whether we are actually subject to an autonomous technology.

2.2.3 Social constructivism

We have seen that there is a clear interaction between technology and society. The question is what the exact nature of this interaction is. David Kaplan writes 'any technology has a social meaning relative to its use and context'.[30] A car, he points out, is both a means of transport and a status symbol. It enables social reach. Critical readings of technology evaluate its effect in measures of social justice and human happiness. In its hard version, social constructivism sees society driving technology, in its soft version, society and technology are co-determined.

Trevor Pinch and Wiebe Bijker were amongst the early proponents of social constructivism. In their extensive review of the literature they claim that 'all knowledge and all knowledge claims are to be treated as being socially constructed',[31] 'there is widespread agreement that scientific knowledge can be and indeed has been shown to be thoroughly socially constituted', even to the extent that according to Pinch and Bijker 'there is nothing epistemologically special about the nature of scientific knowledge' and scientific and technological knowledge is 'a sociological task not an epistemological one'. They quote Mayr to the effect that 'science and technology are themselves socially produced in a variety of social circumstances'. As exemplars they quote the dumping of war supplies of phenol making Bakelite competitive against celluloid, Callon's study of electric vehicle development in France, Noble's study of numerically controlled machine tool development, and Lazonick's study of the self-acting mule. Whilst these studies undoubtedly demonstrate the effect of social variables on the development and deployment of technology, they by no means suffice to support Pinch and Bijker's exaggerated claims for a social constructivist theory of scientific knowledge and technological possibility. The natural world generates the artefacts of science and technology, and the natural world is objective, not a social construct. However, social constructivism, even in its weaker form of claiming interaction between society and technology, does offer a valid challenge to the determinist theory of technology.

Langdon Winner is strongly critical of the social constructivist theory of technology. In his classic article 'Social Constructivism: Opening the Black Box and Finding it Empty',[32] his main critique is that, in their determination to attack determinism of technology, the social constructivists may have a programme but lack any theory. He claims that 'all their emphasis is on specific cases and how they illuminate the repeated hypothesis that technologies are socially constructed' whilst they 'sidestep questions that require moral and political argument'. He compares Pinch and Bijker unfavourably to Marx, Heidegger, Mumford, Ellul and Illich, all of whom he claims offer an ethical valuation of technology. Winner wants to do the same. For him 'the key question is not how technology is constructed, but how to come to terms with ways in which our technology centred world might be reconstructed... inspired by democratic and ecological principles'. In this Winner shares an agenda with Andrew Feenberg: neither wants a deterministic technology, including a socially constructed one, because both want to implement political control of technology and so require it to be politically contingent.

2.2.4 Technocracy

Andrew Feenberg writes that 'political democracy is largely overshadowed by the enormous power wielded by the masters of technical systems: corporate and military leaders and professional associations of groups such as physicians and engineers'.[33] The novelist C P Snow captured the emergence of 'technocracy' in his series of novels 'Strangers and Brothers' and 'The Two Cultures', coining his memorable phrase 'the corridors of power' and lecturing on the gulf between scientists and literary intellectuals, with scientists assuming social control due to the sheer power of the military and civil technology they were able to deploy, and the popular economic benefits of technological production.

At the time such fears seemed real, but unpredictably, social structures and economic market parameters have dethroned the technician of modernity and enthroned the celebrity of post-modernity, substituting image for content, and 'celebrocracy' for 'technocracy'. Markets commoditise, and competitive markets have commoditised science and technology, diminishing their social power. This could well result from the increased private market orientation of technology, compared to its state organisation at the time C P Snow wrote and lectured from his role in government, responsible for science and technology.

2.2.5 Utopia and dystopia

The nineteenth century saw a huge development in science and technology which caught the popular imagination and engendered a positive expectation of ever increasing prosperity from humanity's new understanding of the science of the universe and control over the technology derived from it. Robert Scharff compares the response of nineteenth century Auguste Comte and twentieth century Martin Heidegger to 'technoscience'[34]: Comte celebrated it and welcomed its ultimate fulfilment, Heidegger critiqued it, profiling technology as dystopia, a view developed further by Herbert Marcuse, Hans Jonas and Alfred Borgmann.

2.2.5.1 Heidegger

Robert Scharff and Val Dusek write that 'Martin Heidegger's interpretation of technology...is probably the single most influential position in the field'.[35] Andrew Feenberg agrees that 'Heidegger is no doubt the most influential philosopher of technology in this (twentieth) century'.[36] Heidegger's core article is 'The Question Concerning Technology' written in 1954.[37] Heidegger's text is abstruse and inaccessible; whether as a result of translation from the German, or due to the predilection of philosophers for such style, is not certain. Typical Heidegger analysis is that 'we should like to prepare a free relationship to (technology); the relationship will be free if it opens our human existence to the essence of technology; when we can respond to this essence we shall be able to experience the technological within its own bounds; everywhere we remain unfree and chained to technology, technology is a way of revealing, technology (is)...where revealing and unconcealment take place, where truth happens'.

An attempt to interpret this into meaningful accessible language (ironically, for Heidegger himself on the question of language see below), might be

1. Traditional technology reveals or 'unconceals!' energy (or products) from nature, bringing forth its latent pregnant potential – the main idea is some kind of incarnation, or specific definition of nature as production potential
2. Modern technology is more aggressive and 'challenges' or 'sets upon' nature – 'a revealing that challenges' – the idea here seems to be that modern technology is a demanding exploitation of nature

3. Technology redefines nature as a resource for exploitation – a standby – Heidegger's 'standing reserve'
4. Technology then redefines humanity as a mere resource – both consumers and workers are resources to the system – man is 'enframed' in the system process – this is the ultimate danger

Heidegger's famous phrase 'only a god can save us now' taken from his interview with Der Spiegel in 1966 published posthumously in 1976,[38] was widely taken to refer to the devastating potential of technological power, exemplified by nuclear power, which had become autonomous and threatening to the very human race which had developed and deployed it. In the wake of Hiroshima and Nagasaki, it is easy to see the validity of such a view. Heidegger's reply to Spiegel was 'If I may answer briefly, and perhaps clumsily, but after long reflection: philosophy will be unable to effect any immediate change in the current state of the world. This is true not only of philosophy but of all purely human reflection and endeavour. Only a god can save us. The only possibility available to us is that by thinking and poetizing we prepare a readiness for the appearance of a god, or for the absence of a god in [our] decline, insofar as in view of the absent god we are in a state of decline. So the crisis is not only from technology but in philosophy also'. When asked what would then replace philosophy, Heidegger replied 'cybernetics'.

However, there are great difficulties with this analysis, requiring a comprehensive critique of Heidegger. In his preference for traditional technology over modern technology, exemplified by his fond description of rural bridges, Heidegger is simply nostalgic, a nostalgia which Alfred Borgmann rejects. It is a very partial analysis of former low technology societies where all was not at all bliss. But the greatest problem with Heidegger is his mute response to the Nazi regime.[39] Heidegger faced very great and valid criticism for his acceptance of the role of rector of Freiburg university which was in the Nazis' gift. Here is a man who fails to see the danger in Nazi national socialism, but writes with great urgency about the danger he *is* able to see in modern technology and in America, a country he constantly criticised but refused to visit. I submit that this is a psychological distortion in Heidegger. One failure to identify a very real danger is over-compensated for by an exaggerated identification of another lesser danger. Thus he lectured in 1949 that 'Agriculture is now the mechanised food industry, in essence the same as the manufacture of corpses in gas chambers and extermination camps, the same as the blockade and starvation of nations, the same as the production of hydrogen bombs'.

Such a comparison can only too easily be read as a quasi-excuse for the Nazi terror by diminution, and is therefore repudiated. A valid hypothesis can also be offered that America attracted Heidegger's strong critique because it became the global techno-power that he wanted Germany to become. His pervasive racism continues remarkably in his 1966 interview with Der Spiegel, where he suggests that French philosophers found the French language inadequate for expression of philosophy and preferred the German language for this!, viz, 'I am thinking of the special inner kinship between the German language and the language of the Greeks and their thought. This is something that the French confirm for me again and again today. When they begin to think, they speak German. They assure [me] that they do not succeed with their own language.' Here is unquestioning, unremitting, unapologetic nationalism.

However, if we interpret a clearer simpler unencumbered version of Heidegger, we can agree that technology can have a tendency to redefine our world: so that both nature and human beings are simply resources for technological production which can potentially drive the process. A tree is no longer a tree per se, but timber for furniture production. A landscape is now a deposit of iron ore, coal, oil, or some other mineral. And the human being is simply a labour resource. This threatens to take the soul out of both nature and humanity. Religious and secular environmental views counter this interpretation by insisting on the independent ontology of both nature and humanity. However, reductionist and complete definitions of nature and humanity are not necessarily mutually exclusive. Definitions can be explicitly partial without negating wider total definitions. For some purposes, for example in arranging a social occasion where seats have to be ordered, I am content to be defined as a unit of one, without this detracting from my full humanity in a more total definition. For economic purposes, I may be considered either as a resource or a consumer, without this denying my full humanity. Heidegger's point is therefore taken, but it does not necessarily convey the reductionist import he suggests.

Heidegger supposes that the artefact of technology precedes both nature and humanity. This is an unusual claim for any artefact. Reification is a process which is usually conceived to follow human presence and action, as in the artefact of market. Heidegger gives no explanation of how technology might precede humanity. Its constituent elements of nature and scientific process are objective and prior to humanity, but the reconfiguration of nature as material and scientific process is essentially a human creation. So any reification of the artefact of technology is consequential to and contingent on humanity.

Subjective perspectives determine how nature and humanity are defined and regarded. The choice of cynical or optimistic interpretation is entirely arbitrary. There is some parallel here in how human labour is regarded in economic theory. To the neo-classical economist, labour is a cost of production and should be minimised. Wages were therefore argued down in the Great Depression. Keynes showed how labour was not only a cost of production, but also the source of demand in the economy, and he therefore argued for effective real wages in the same depression. Similarly, human labour can be regarded cynically as a mere dehumanised resource in technological production, but the same human labour is also the consumer of the product, and enjoys its benefits and the increased leisure time enhanced productivity can offer. The exception to this is where the producer is not also the consumer, as in the inequality of early periods of technological production chronicled so piercingly in Engel's 'Condition of the Working Class in England 1844', or apparent in the phenomenon of low-wage Chinese production for American and European market consumption today. Otherwise, technology harnesses humanity as a labour resource, but also supplies humanity as consumer. Technology is not an independent prior artefact but is contingent on humanity. It cannot therefore precede and control humanity. So while Heidegger accuses technology of master control status, optimists see technology as the tool that has elevated the otherwise weak and impoverished human species to the status of lord of nature. A more subtle analysis than either of these is needed and is explored later.

2.2.5.2 Herbert Marcuse

Marcuse carries the cynical dystopian interpretation further. A selection of quotations from his chapter 'New Forms of Social Control' in his 1964 book 'One Dimensional Man'[40] conveys the theme of his dystopian view of technology. It is, he claims, 'the suppression of individuality in the mechanisation of socially necessary but painful performances'. In this process, 'the rights and liberties which were such vital factors in the origins and earlier stages of industrial society... are losing their traditional rationale and content'. Whilst 'freedom from want is becoming a real possibility' the downside is that 'the liberties which pertain to a state of lower productivity are losing their former content'. In post-modern form he states that 'to impose Reason upon an entire society is a paradoxical and scandalous idea' and the result is that 'in the contemporary period the technological controls appear to be the very embodiment of Reason for the benefit of social groups and interests – to such an extent that all contradiction seems irrational and all counteraction impossible'. 'Today', he claims, 'this private space has been invaded and whittled down by technological reality. Mass production and mass distribution claim the entire individual ... advanced industrial society silences and reconciles the opposition ... the efficiency of the system blunts the individual's recognition that it contains no facts which do not contain the repressive power of the whole'. There is much more in the same vein. Marcuse may be correct in pointing out that the pleasurable consumption afforded by technological production does obscure the realities of the system, from both consumer and producer, who participate in it but are alienated from it. However, this can be overcome by some basic educational explanation of how the system works. Marcuse's cynicism typifies the general tendency of European intellectual thought compared to more positive American thought, and indeed this distinction does arise in the philosophy of technology with more empiricist American contributions – reviewed below. Cynicism can be great sport, but one might ask where Marcuse was living or thinking of when he portrays a previous era in such glowing terms when rights and liberties were more respected than in the technological age, in which alternatives are available to humanity. Serfs living under feudalism are unlikely to have agreed with his evaluation of their predicament.

2.2.5.3 Hans Jonas

Jonas adds his voice to the dystopian perspective, claiming that whilst 'technology adds to the very objectives of human desires', yet 'nobility has been exchanged for utility'[41] and 'the technological syndrome has brought about a thorough socialising of the theoretical realm, enlisting it in the service of common need ... elevating homo faber to the essential aspect of man ... elevating power to the position of dominant and interminable goal'. He is right when he points out that 'an intricate web of reciprocity has been part of modern technology' and this aspect of his thinking is developed further as the conclusion to this book. However his conclusion that we live in a 'novel state of determinism – Man may have become more powerful; men very probably the opposite, enmeshed as they are in more dependencies than ever before' and his warning against 'the quasi apocalyptic prospects of the technological tide – disaster must be averted' would hardly convince the average citizen, even in the bad old days of 1964 when Jonas wrote these lines!

2.2.5.4 Alfred Borgmann

Borgmann offers a rather milder critique.[42] He worries about the imbalance between means and ends in the deployment of modern technology. He warns about the 'suppression of the value question' and 'the enslavement of humankind to its own invention'. Andrew Feenberg is dismissive of Borgmann's concerns, saying in particular that he offers no criteria for constructive reform, and that in his critique there is 'an element of ingratitude'.

2.2.6 New conceptualisations of technology – American empiricist phenomenology

As technology brought increasing economic benefits to the developed world in the second half of the twentieth century through its impact on productivity, raising per capita output and therefore standards of living; and as the mega-catastrophes of nuclear Armageddon and a fertiliser/pesticide wasteland were avoided, the ultra-dystopian cynicism of Heidegger, Marcuse and Jonas lost apparent relevance. They were also critiqued as a top-down theory determined prior to consideration of the data and often with a strong political bias. In 'American Philosophy of Technology – the Empirical Turn', Hans Achterhuis and a team of philosophers of technology at the University of Twente in Holland review the alternative perspective developed more recently by six American philosophers: Alfred Borgmann, Hubert Dreyfus, Andrew Feenberg, Donna Haraway, Don Ihde, and Langdon Winner.[43] As Achterhuis points out, the classical 'godfathers' of the philosophy of technology were strongly dystopian, whereas these recent American writers are more optimistic. However, the shift in methodology is more significant than the switch to optimistic mode. Achterhuis characterises the new thinking as (1) empirically oriented, in that the data of the phenomenon of technology is researched and profiled before macro philosophies of technology are developed (2) technology is then perceived, not as autonomous but subject to many social forces with which it co-evolves. There is he says 'no inherent logic pushing technology and society relentlessly in the direction of greater efficiency or uniformity'. Alfred Borgmann 'laments the transformation of culture at the hands of technology', Hubert Dreyfus challenges the ability of the technology of artificial intelligence to replace human intelligence, Donna

Haraway presents her cyborg ontology of techno-humanity, and Andrew Feenberg and Langdon Winner argue for democratisation of technology.

It is Don Ihde's work which most clearly presents the methodology and philosophy of 'phenomenology', although Feenberg's work, for example in his famous review of the French Minitel system, also deploys the phenomenological method and as with Ihde arrives at more positive conclusions about technology. In his 1990 book 'Technology and the Lifeworld',[44] Ihde interprets the Biblical Garden of Eden myth to address the question of whether humanity can exist naked, that is, non-technologically, humanity's vulnerability requiring that this should be in a garden. He points out that we still retain direct sense perception and direct bodily motility,[45] but that we see 'an artificially aided perception of nature'[46] through instruments. We see previously unknown worlds that 'we could not see when naked'. This does bring a 'decision burden' to humanity which can no longer live as naively as in the garden. In these ways, Ihde's interpretation of the myth is innovative and compelling. Instrumentation technology is a key Kuhnian paradigm change, with instrumentation relating to humanity through embodiment, interpretative hermeneutics, or 'other' alterity. The instrument is rarely totally transparent, but effects some transformation of perception. Clocks mediate a time culture, and compasses a space-location culture. Shape is physical, but colour is metaphysical and subjective: this maps onto technology which is materialist, and lifeworld which is metaphysical, potentially independent of technology, but interactive with it. For Ihde, the reification of technology is 'over-metaphysical'.[47] His philosophy of technology is 'rigorously relativistic', so that he denies that technologies are neutral since that would render them non-relativistic 'objects in themselves'. He wants to preserve a dynamic interaction between humanity and technology and offers a structuralist account, though not proposing what this structure is.

Peter-Paul Verbeek reviews Ihde's phenomenological philosophy of technology. He shows that the central questions in Don Ihde's philosophy of technology are[48]

- what role does technology play in everyday human experience?
- how do technological artefacts affect people's existence and their relations with the world?
- in particular, how do instruments produce and transform human knowledge?

This is, says Verbeek, 'a perspective that seeks closer contact with concrete technologies'. It is data driven, deriving theory from phenomenological data, rather than theory driven. In particular the philosophy of phenomenology rejects subject/object duality. Specifically it rejects the human/world duality which typifies earlier utopian and dystopian philosophy of technology. Technology therefore is a contingent intentionality. Consciousness and perception are 'of something'. Instead of following Heidegger in 'reducing technological artefacts to the technological form of world disclosure that makes them possible', Ihde asks 'what form of world disclosure is made possible by technological artefacts'.[49] Also contrary to Heidegger, 'technology can even allow the world to manifest itself in new ways...technologies transform perception differently'.[50] The symbiosis between humanity and technology is presented – 'Once taken into praxis one can speak not of technologies in themselves, but as the active relational pair, human-technology....technologies are indissolubly linked with humans-in-culture...technologies have no essence; they are only what they are in their use'.[51] The important conclusion of this is that 'multistability', whereby technologies can adopt alternative use-forms, makes the substantivist position untenable. 'Technology cannot be understood as an independent power that holds culture in its grip, for its form is ambiguous; it becomes what it is only in the context of culture'.[52]

But Verbeek counters, 'technology is as little neutral as it is determining': technological intentionality, choice and action is required. He cites examples of writing technology, for example, the fountain pen, typewriter, and word processor, and discusses their effect on writing style. Technologies create a 'decision burden'. He concludes that 'Ihde's arguments concerning the coming about of a pluricultural life-world, and the increasing contingency and decision burden, weaken the instrumentalist conception of technology'.[53] Asking 'Is the cultural relation to technology multistable, or do technologies have a culture changing power?' he concludes 'The predictions of analytical uniformity (Marcuse) of the victory of technique (Ellul) and even of the sheer world of calculative thought (Heidegger) are wrong. There will be diversity, even enhanced diversity, within the ensemble of technologies and their multiple ambiguities, in the near future. Technical culture does not develop in the direction of one-dimensionality, calculativity and uniformity, but rather in the direction of plurality. Technology does not create one single way of disclosing reality – the 'technological way of revealing' – rather it fosters the proliferation of different ways of seeing within our culture'.[54]

This is both a more practical and more realistic way of perceiving technology and analysing its interaction with nature and humanity. It undergirds Ihde's work in instrumentation technology and Feenberg's in analysing the social value of Internet technology.

2.3 Moderating technology – taming the beast?

If technology is not instrumental and therefore neutral, but does have some element of independent ontology, then how can its potentially negative effects be moderated and managed? The answers proposed to this are variously spiritual, social and conceptual.

2.3.1 Heidegger's 'saving power'

Heidegger proposes an extremely vague spiritual solution to the threat of technology. He apparently believes in some 'saving power'. His writing on this salvation in 'The Question of Technology' is Heidegger at his most mystical[55] – viz

> 'The essence of modern technology starts man acting upon the way of that revealing through which the actual everywhere more or less distinctly becomes standing reserve. We shall call the sending that gathers 'destining'. It is from this destining that the essence of all history is determined. Enframing is an ordaining of destining. For man becomes truly free only insofar as he belongs to the realm of destining and so becomes one who listens though not one who simply obeys. Man is endangered by destining. The destining of revealing is in itself not just any danger but the danger. Enframing blocks the shining-forth and holding sway of truth. Technology is not demonic, but its essence is mysterious. But where danger is, grows the saving power. For we have said that in technology's essence roots and thrives the saving power. As the essencing of technology, enframing is what endures. The granting that sends one way or another into revealing is as such the saving power. For the saving power lets man see and enter into the highest dignity of his essence. The irresistibility of ordering and the restraint of the saving power draw past each other like the paths of two stars in the course of the heavens. We look into the danger and see the growth of the saving power. Once the revealing that brings forth truth into the splendour of radiant appearance was also called techne.'

Heidegger's expression here is so mystical as to be meaningless. The essence of technology he claims is mysterious. So is Heidegger's saving power, unless it connects to his proposal that humanity is free from technology threat once humanity is aware of the artefact of technology. This could be interpreted to mean that freedom comes from understanding what is going on regarding technology and humanity, but such an interpretation, though promising, is undoubtedly too simple as an interpretation of Heidegger.

2.3.2 Ellul and Borgmann – faith and focal things

Ellul and Borgmann are both Christians and propose a corrective to a potentially dominating technology in determined emphasis on non-material spiritual elements of a holistic humanity. For Ellul this is the simple faith itself, but Borgmann is more explicit and specific in his advocacy of 'focal things and practices'.[56] Thus he takes Heidegger's beloved jug and offers multiple holistic interpretations of this artefact. It can be used to hold, to offer, to pour or to give. It gathers clay, wine, sun and sky. It refreshes humanity and offers a divine libation. He suggests that 'the orienting force of inconspicuous humble simple things' can offer humanity an orienting force to counter or at least balance technology. Focal practices include running (unencumbered by technology apart from the advanced trainers?), or a family or community meal.

2.3.3 Habermas and Feenberg – democratisation

Habermas and Feenberg offer stronger remedy. Having concluded that 'this thesis of the autonomous character of technical development is not correct',[57] Habermas is hopeful that 'the pessimistic assertion that technology excludes democracy is just as untenable'. 'Our problem can be stated as one of the relation of technology and democracy'. Similarly Feenberg argues that 'technological determinism is refuted by historical and sociological arguments' so that 'democracy can be extended beyond its traditional bounds into the technically mediated domains of social life'.[58]

Whilst Feenberg gives examples of popular activism in the use of technology, particularly Internet technologies, neither he nor Habermas sets out how the democratisation of technology could operate more comprehensively.

There are a number of problems with the proposal for democratic control of technology.

A detailed specification of the proposed democratic process is required; for example, how the franchise is to be defined, whether the process is by referenda or by representative democracy, whether the threshold for a positive vote is 50 per cent, and whether this is of those voting or of those entitled to vote. All of these are flawed in some way. For example 51 per cent democracy offers potential suppression of the 49 per cent in disagreement. Referenda are difficult to formulate meaningfully: for example, how would the UK have proposed a referendum on the Lisbon Treaty for EU governmental procedures – would an all or nothing 'yes' or 'no' vote be the only possibility? How meaningful would this be? How would a complex set of procedures – where someone could always object to some sub-clause – ever get accepted in a referendum and so how would society govern itself?

Activism is one alternative that Feenberg appears to favour in his examples, but this can easily lead to control by very small minority groups. Gandhi successfully promoted traditional homespun technology as part of his political agenda, but he did not calculate the constraint this would impose on the standard of living in rural India, nor did he explain this to the franchise. There have always been Luddites and opinions opposing technology, but their views are often very partial and, as Anthony Crosland pointed out in debate on environmental policies with E J Mishan, often very privileged. Other writers have pointed out that the franchise would need to be sufficiently well educated in the technology to be voted on, and this requirement becomes increasingly difficult to achieve as the technology becomes ever more complex.

Contemporary popular activism focuses on technologies such as genetically modified (GM) food crops, nuclear power, fossil fuel power generation, stem-cell research. It might be easy for western consumers who feel assured of sufficient food supply to vote against GM crops, without taking into consideration the large numbers of people in developing countries subsisting on an inadequate diet. Nuclear power attracts ready opposition, but provides zero-emission electricity such that a nation like France, relying for over 80 per cent of its electricity on nuclear generation, enjoys the longest life expectancy of any comparable country. The Green lobby in Germany secured government action to close 21 gigawatts (GW) of nuclear power generating capacity, but in the early 2000s this led to substantially increased coal-power generation which raised emissions, as was fully and publicly reported on the major German power-generators' web sites. Before becoming part of the coalition government, the Liberal Democratic party in the UK opposed new nuclear-power generation capacity and new coal-power generation that is not accompanied by carbon sequestration technology. However, it did this (1) without saying how it proposed to reduce or generate the 350 terawatt-hours (TWHrs) of power consumed each year in the UK economy, (2) what the consumer kilowatt-hour (KWHr) unit cost is likely to be of power generated by wind farms or clean coal, and indeed (3) whether such power is available in the medium term. These examples demonstrate that technology decisions are complex and are probably best left to representative democracy advised by technical experts, which is the current system. There are already a significant number of controls for any new technology to surmount before it can be marketed.

How is a democratic resolution to be policed? In many cases, a single nation's decision to ban a technology is meaningless in the face of its adoption in other countries which are then free to market products derived from this technology on the world market. If western countries ban stem-cell research, then societies with more pragmatic ethics may well develop the technology.

It is doubtful, therefore, that political democratisation of technology could be made to work effectively. The question then arises as to whether the market offers any element of democratisation of technology? Certainly the privatisation of many macro-technology markets, from power generation to utilities and aerospace, has led to more competitive, fragmented and politically less powerful technology bases. Consumer power can be significant: boycotts of technologies or goods and services produced by unacceptable technologies are a potential democratic process. Packaging technology, for example, has recently been substantially affected in this way.

3
A Comprehensive Systems Network Philosophy of Technology

Having set out and critiqued the current body of academic thinking on the philosophy of technology, a more comprehensive systems network concept for the philosophy of technology is now developed. The process adopted for this is to first boldly state the hypothesis of the model, then to examine its constituent entities and relationships in detail, thus generating a critique to reformulate the model in subsequent iterations.

Foundation assumptions of the model are

- technology is defined as the cognitive human reconfiguration of natural materials and natural scientific processes
- technology is therefore wider in capability than nature since it deploys configurations of natural elements where the configurations are not necessarily found in nature
- nevertheless technology is limited to possible reconfigurations of natural materials and processes
- technology is therefore entirely contingent on natural materials, natural scientific processes and human intention, decision and action
- technology is therefore not uniquely determinist, not autonomous, and as an artefact is neither independent of nor preceding humanity
- technology harnesses science and impacts human artefacts of economy and society, and thereby humanity itself through productivity
- technology is located in an extensive systems network of nature, humanity, productivity, the market economy, and society
- the artefacts of technology, productivity, market economy and society interact and are co-determined in the systems network
- all interactions in the systems network are essentially multivariate and not bivariate
- the symbiosis which is evident between humanity and technology is best understood through this systems network and not directly bilaterally

- the systems network responds to exogenous change which can occur at any of its nodes, and the system responds and converges to a new equilibrium
- the systems network is extremely dynamic with regular movement and systems-wide evolution
- whilst nature and therefore technology are constrained, the result of their network systems interaction on human economic systems and social structures is not readily susceptible to prediction
- Heidegger was right in that understanding this complex systems network for the philosophy of technology allows humanity to be free from any perceived threat from an independent dominating technology – he simply failed to explicate any such model as a liberating understanding

The systems network hypothesis is set out in the following diagram in Figure 3.1. (facing page) Issues are noted against each entity and relationship in the model for subsequent exposition.

Figure 3.1 sets technology in a structural context. This is consistent with the recent more general emphasis on structuralism in philosophy, for example in the philosophy of mathematics. Nature and humanity are shown in red as the two real phenomena, whilst science, technology, productivity, the economy, and society are shown in blue to represent their status as artefact. Some relative importance is recognised in the size of each symbol, with nature being the largest, humanity next, and artefacts the smallest.

This does represent an alternative, more comprehensive hypothesis for a philosophy of technology – or put more generally, for an understanding of technology – than has so far been offered in the literature. The claim is that this model does embrace various strands of thought presented in the literature in a more disaggregate way, and then defines an overall systems network in which these issues can be usefully located. Without such an overall structure, the effects of any more limited interactions are difficult to analyse. The model embraces various bivariate relationships but replaces them by insisting on a wider multivariate analysis.

It is therefore necessary to justify the model and its underlying concepts in order to establish its claim to such a status. We do this by working through

i) the model's assumptions
ii) the model's entities
iii) the interactions between the model's entities.

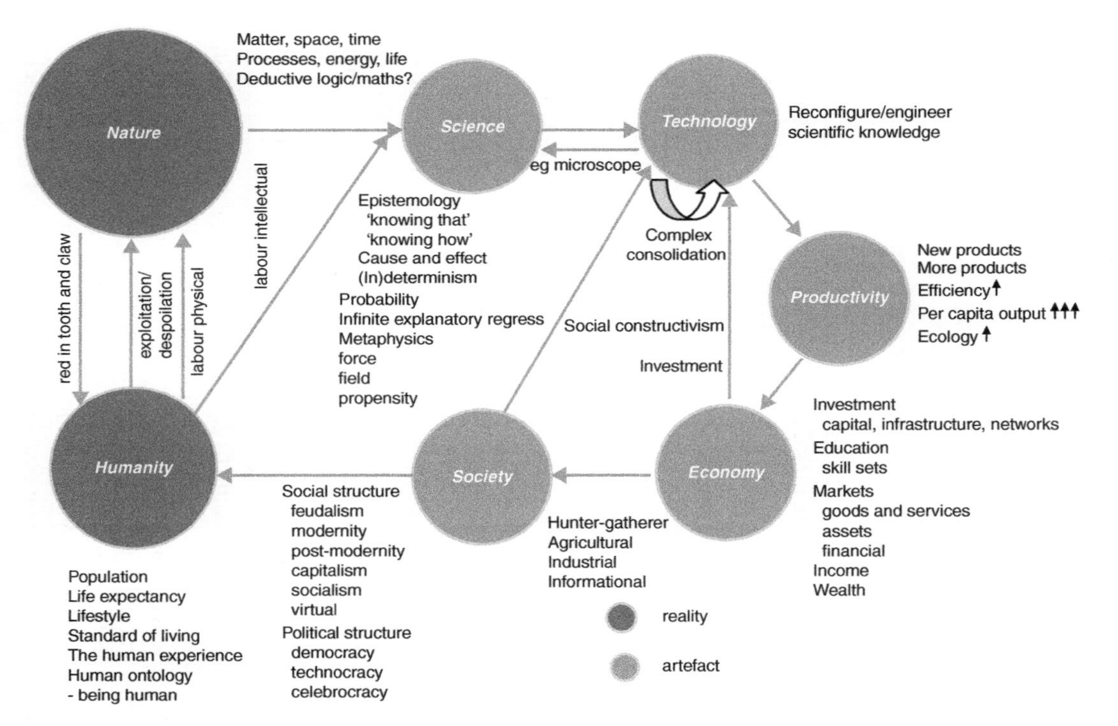

Figure 3.1 The human symbiosis – nature mediated through science and technology via productivity

3.1 The model's assumptions

3.1.1 Technology is defined as the human reconfiguration of natural materials and natural scientific processes

This definition is preferred to the other definitions developed in the literature of rationality, human enhancement, and instrumentalism. Rationality is a claim about nature; that deductive logic is objective and somehow located in nature. It is therefore a claim about nature rather than technology itself. By defining technology as rationality we are required to primarily study rationality as a route to understanding technology. Technology certainly incorporates rationality, but it is something more than that. Technology can also be irrational in two ways. Firstly its application may follow subjective whim rather than logical rational analysis. Secondly, its functionality may in fact not be totally rationally understood. This applies to a lot of technologies. The exact way in which television waves convey their information through the atmosphere is not totally understood. We simply know it works and we use it. Keewok Lee claims that most technology was of this type up until 1850[59], that is, we had discovered and applied a process without understanding how it worked, whereas after that date we focussed on synthesising technologies from an understanding of their constituent scientific rationality. Lee's observation of the shift in emphasis is correct, but the shift is by no means total. We use the measured principles of gravitational attraction constantly in the technology we engineer, but we do not know how gravitational force works. The same is true of electromagnetic field and of the probability distributions we have to engineer technology to cope with. Our rational explanations are limited and always subject to infinite regress, an intellectual challenge we examine further in a later section.

Similarly, human enhancement is not a definition of technology at all. Technology may play the role of enhancing human power, but this does not define technology for us. A car may convey its human occupants from location A to location B, but 'that which conveys human occupants from location A to location B' is not an adequate definition of a car. It would not help you to build one, although it may spur you to seek to define one so that you could build one. Merely saying that technology is the enhancement of human power fails to say 'what' it is that is doing the enhancing. It doesn't define this 'what' at all. Furthermore as we have pointed out above, technology can actually reduce human power. This is true of artificial intelligence. Computers can only just play chess to a standard which matches human master chess champions. And in other areas of conceptualisation, Dreyfus' work has shown artificial intelligence to be inferior to the capability of the human brain. Artificial intelligence, or more general information and computing technology, can extend human power through extensive number crunching and algorithms, but reduces it at other levels of conceptualisation. It therefore becomes necessary to amend the definition to say that technology enhances human power in some dimensions, whilst reducing it in others. This hardly leaves us with a meaningful definition.

We have said why cyborg terminology is not a definition of technology: it is a definition of humanity rather than of technology. Like the definition of technology as human enhancement, it does not say 'what' it is which is rendering humans into cyborgs. A formula that cyborgs minus humanity = technology is not meaningful or workable.

Instrumentalism is also an inadequate definition of technology because technology outcomes can surprise the human developer, contradict her intention, and more often simply advance into application unnoticed. Instrumentalism as a definition fails to allow for any independent status for technology. It predefines technology as excluding such independence, and so closes down the philosophical questioning of technology which is so important. It therefore has to be rejected as a definition of technology.

Philosophy has to address human questions in terms familiar to human experience and understanding. The phenomenon of technology is commonly understood and experienced as the engineering of natural materials and scientific processes into new configurations which are then applied into infrastructures, products, and services. It is this phenomenon of technology which philosophy has to consider and philosophise about. Philosophy should not invent its own exclusive definition of commonly understood terms, as it has with the definition of a gene in the philosophy of biology, but should accept the challenge and task of considering the essence and implications of commonly defined phenomena. Only in this way can philosophy hope to harmonise in symbiosis with humanity, a symbiosis it has all too often lost sight of or interest in.

The definition of technology adopted in this model is therefore proposed and defended as true to common understanding, and meaningful for further philosophical reflection. It is that technology is

'The cognitive human reconfiguration of natural materials and processes'

3.1.2 Technology is therefore wider in capability than nature since it deploys configurations of natural elements where the configurations are not necessarily found in nature

This is on the one hand evident, but on the other hand curious and interesting. We normally think that if A leads to B, and if B is entirely dependent on A, then B cannot be greater than A. B is seen as a subset of A. However, once we allow for the process of reconfiguration of A, then B can indeed be greater than A. A is a toolkit and B the wide range of outcomes possible from A. Nature is therefore a toolkit for technology to build its extensive outcomes which then reach beyond nature itself. There are for example no plastics found in nature, but reconfiguration of natural materials, in this case oil and hydrocarbons, with scientific processes also distilled from nature, produces plastics which are beyond nature. The same is true for many inorganic chemicals and for base-element metals which are only found as ores in nature. Even where technology has outputs which are found in nature, for example CO_2, technology can produce proportionally greater volumes than are found in the initial balance within nature. Using only nature's inputs, its materials and processes, technology can therefore not only generate outcomes beyond nature, but can also change nature itself. This introduces a dynamic interactive process in that

nature = nature's materials + nature's scientific processes + cognitive human reconfiguration
= technology + production
= infrastructures + products + services + emissions → **nature**

This has huge implications for ecology. Nature is reconstituted with each round of the technology process, amended each time with new materials, an excess of some existing materials, and depletion of other materials. The same human cognition which harnessed nature's materials and processes in the first round now has to exercise responsibility and take creative action for the second round outcomes in the redefinition of nature it has caused. The process needs responsible management, a responsibility which starts with awareness of the technology process and its impact.

3.1.3 Nevertheless technology is limited to possible reconfigurations of natural materials and processes

Again this seems at face value an obvious statement. But it is an important assumption for the model because it does set limits and constraints on technology's possibilities. Technology cannot simply do everything. It cannot for example strike a match on jelly, well, not the match and jelly we currently know and love. This in turn has two implications. On the one hand, the fear that technology may stretch to infinite capability, and that this will threaten both nature and humanity, may be exaggerated, although it is admitted that the threat of nuclear technology is very real. On the other hand, the supreme serene confidence that technology will resolve every need and crisis is also misplaced. The easy conviction that, because we have seen technology make huge strides beyond any expectations in the last 200 years, therefore it will soon come up with handy solutions to our energy and emissions crises is over confident, specious, hasty and presumptuous, and therefore dismissive of a real problem we face.

3.1.4 Technology is therefore entirely contingent on natural materials, natural scientific processes and human intention, decision and action

We have criticised the instrumental definition of technology because it presumes human control of technology and thereby excludes the possibility of an independent autonomous threatening technology. At first glance this definition does the same. If technology is contingent on human intention then we have the same constraint on our understanding of technology as we get from the instrumental definition. It is therefore important to define in what way technology may be contingent on human intention, decision and action. As we have said, technology developments can and have resulted in outcomes which have surprised and even contradicted the intention of human agency. Nevertheless, whilst technology can generate unexpected and controversial outcomes, it still neither develops itself, nor puts itself into application. It does need human agency. Whether that human agency in all cases retains total control over every aspect of the technology process is the leading question we address later. Some human intentionality is therefore necessary to the definition of technology. Human agency has to initiate the process both of technology development and of its application, although the outcome may contradict the original human intent. We have to distinguish and differentiate between human intentionality in activating the technology process, and human intentionality regarding the outcome of the process. Intentionality is stronger in the activation than in the outcome of technology. This nuances our understanding of the role of intentionality in the technology process, and leaves room for consideration of some exogenous element in the nature of technology.

3.1.5 Technology is therefore not uniquely determinist, not autonomous, and as an artefact is neither independent of nor preceding humanity

This assumption is also partly a conclusion. It therefore has the status of a 'null hypothesis' at this stage of the discussion. The previous assumptions give ground for this further assumption that technology cannot determine humanity. We shall see as the discussion progresses, that, whilst this is theoretically the case, and practically also possible, there are also practical details which can and often do inhibit the complete control over technology which is both theoretically and practically available to humanity. There are therefore three dimensions against which any claim has to be checked, in this case the claim for autonomy and determinism of technology, the dimensions of a priori theory, real time practicality, and ex post reality.

3.1.6 Technology harnesses science and impacts human artefacts of economy and society, and thereby humanity itself through productivity

This assumption directly challenges Heidegger's view that technology exists metaphysically prior to and independently of humanity. Whilst mathematics appears to be discovered rather than invented, and so has metaphysical status, and whilst this may also be true of deductive logic, technology is an engineered reality. Under the definitions above, it therefore cannot precede humanity, although it can be a reified artefact once it has been created by humanity. It shares this status with the concept of market. A market does not exist prior to humanity's creation of it, but once it does exist, the market has considerable potential to constrain humanity in ways which humanity, as its original creator, may well not have intended or wanted. Market is in fact a technology. Technology does harness science, but has a wider portfolio than that offered to it by science, since technology can also harness phenomena which are not scientifically understood and have not been fed to it from science. We suggest a resolution to this dilemma by defining science both as 'knowing that' and 'knowing how' – see below.

What is important in this assumption of the model is that technology is stated to have impact on other human artefacts, such as the economy and human social structures. It does this through its effect on production methodology and on productivity. Technology does require specific production facilities and methodologies to produce the infrastructures, products and services which it renders available. Factories are needed, and these in turn need mass workforces which together lead the process of urbanisation. Urbanisation in turn has a huge impact on human living, creating anonymous, private and often lonely lifestyles, all of which in their turn have yet further consequences. Technology also drives up productivity and this, as we shall see in a later chapter, massively redefines the human life experience, creating the consumer society, a huge ecological impact, and raising standards of living and sustainable population numbers.

3.1.7 Technology is located in an extensive systems network of nature, humanity, productivity, the market economy, and society

3.1.8 The artefacts of technology, productivity, market economy and society interact and are co-determined in the systems network

3.1.9 All interactions in the systems network are essentially multivariate and not bivariate

Taken together, these three assumptions state that nature, humanity, science, technology, productivity, the economy, and society are co-located in a single systems network and are co-determined within that network structure. It is impossible to understand how any two of these artefacts interact without considering their interaction in terms of this network. Analysis in the academic literature which has sought to do this, often for example seeking to relate technology to social structure, is incomplete. The network systems model of technology is a necessary construct, not a wider optional extra. Attempts at bivariate analysis will be confounded by the multivariate reality.

3.1.10 The symbiosis which is evident between humanity and technology is best understood through this systems network and not directly bilaterally

This is the important specific version of assumption 3.1.9 above. Humanity contributes to the creation of technology, which is also created by nature. Technology then definitely impacts on humanity. But the nature of this interaction cannot be captured and understood simply by considering only humanity and technology, since the way this interaction works is through the other artefacts and variables set out in the network. The network as stated itself may not be complete, but its current definition is sufficiently extensive and challenging and should be the analytic tool by which the symbiosis between humanity and technology is understood.

3.1.11 The systems network responds to exogenous change which can occur at any of its nodes, and the system responds and converges on a new equilibrium

The systems network is proposed as one which has multiple equilibrium points but no necessary trajectory through those equilibria. A plurality of development trajectories is possible, as is a plurality of linked positions for the artefacts in the network. This chimes with Don Ihde's view of the essential plurality of technological worlds.[60] For example, as discussed above, any one technology position can and does co-exist with several different social structures. It is not therefore that a technology uniquely causes a social structure, but that there is a pairing between the two and this pairing of values finds an equilibrium, although the equilibrium may be very transitory as the system dynamic evolves. There can be an exogenous shift in any variable in the network, prompted by a new scientific discovery, or a new financial investment to implement a known technology, or a determination to reduce emissions to the environment, or a preference for virtual network professional working thus re-creating rural communities and reducing transport use, etc. Such exogenous shifts will work their way through the network's equations and establish a new equilibrium position.

3.1.12 The systems network is extremely dynamic with regular movement and systems-wide evolution

The network is live, active and extremely dynamic. The year 1850, which most commentators identify as the beginning of the age of systems technology, is only 160 years ago, but in that 160 years, the state of the human species and the planet it inhabits and uses have changed immensely. Attempts to analyse the network are challenged by how quickly it does move from one equilibrium to another. Assumption

3.1.11 sees the model as a kaleidoscope with inter-connected, emergent patterns of technology, humanity, productivity, economy and society. This assumption sees the same kaleidoscope in almost constant dynamic motion, with new patterns emerging almost before the previous model has settled down. Systems equilibrium is not the stable equilibrium of a cone sitting at rest on its base, nor the unstable equilibrium of a cone pivoted on its apex, but the neutral equilibrium of a cone on its side, constantly rolling along, almost in perpetual motion.

3.1.13 Whilst nature and therefore technology are constrained, the result of their network systems interaction on human economic systems and social structures is not readily susceptible to prediction

Robert Heilbroner thought that technology is predictable.[61] He may be partly right. However, the current state of the network model and all its artefacts and variables would be very difficult to predict from the 1850 point of view. Equally it is genuinely difficult to predict the nature of human life on earth as mediated by the technology of the year 2150. Specifying and estimating the equations of the model is too large a challenge. The best that can be done is to use 'scenario planning' methodologies; we attempt this at the conclusion of the book. This inability to predict future network systems behaviour and outcomes makes it very difficult to critique the system either technically or morally. This may well explain the current widespread ambivalence towards the technology/humanity symbiosis, we really can't say what outcomes are likely.

3.1.14 Heidegger was right in that understanding this complex systems network for the philosophy of technology allows humanity to be free from any perceived threat from an independent dominating technology – he simply failed to explicate any such model as a liberating understanding

This is again closer to a conclusion than to an assumption, so we state it as a working hypothesis. Heidegger, as pointed out above, had huge failings and wrote in an incomprehensible way. Nevertheless, if he can be correctly simplified as saying that we need to understand how technology interacts in symbiosis with humanity, in order to have a free role in this symbiosis, then he was correct. He failed however to suggest any meaningful model as to how these interactions work and so was constrained to lament. The model advanced here seeks to fill that gaping hole left by Heidegger.

3.2 The model's entities

3.2.1 Nature

Nature is drawn as the largest real entity. Most people's definition of nature is a combination of geography, climate, botany and zoology. We think of land and sea, rivers, mountains and valleys. This physical nature is populated by living organisms of plant life and animal life. Trees, flowers, and fauna create verdant scenarios and, even in the desert, crocuses and cactuses survive. Nature teems with fish, birds, lizards, mammals, what the Bible graphically calls 'every living creature according to its kind'. The whole thing is prolific. It contains its own dynamic which we call climate. Wind and rain, sunshine and showers, snowfall and hail, wave and tide, are the commonplace everyday experience of dynamic nature. Less frequent tornados cause havoc. More monumental earthquakes, volcanic eruptions, and tsunamis occur occasionally to devastating effect. 'Nature' is traditionally a benign concept, presumably since to those who have survived its vicissitudes, or inhabit its pleasant places, it is providential, docile, and beautific. But nature can be hostile. The dynamics of its physical elements can kill, as in tsunamis and earthquakes, and its living creatures survive by cruelly devouring each other in the Darwinian food chain.

Beyond the immediate nature of planet Earth, the cosmos of nature includes all the planets, the particle streams, the black holes, and infinity. At the same time, deeply within nature, microscopic atoms, electrons and particles gyrate, collide and reform. Paul Davies gives an accessible account of the planetary and particle world of nature in 'The Goldilocks Enigma'.[62] At the macro level the planets spin on their axes and trace their trajectories. Their universe is ever expanding, driven by an antigravity force which leaves the gravitational force between planets balanced by the mass energy within them. At the micro level, organic cells comprise sub-atomic particles which also spin, apparently at the same speed. Particle physics identifies a wide range of particles – positrons which don't last long since they disappear if they collide with a neutron, muons which only last a few microseconds before changing into electrons or positrons, neutrinos emitted by the sun, billions of which pass through the human body every second. They all have the same level of positive or negative electric charge, or none at all; each has a corresponding antiparticle; all spin with a speed which is a multiple of ½! They are all made from quarks. The Large Hadron Collider at the CERN laboratory near Geneva hopes to identify the Higgs field and the Higgs Boson particle by replicating the moments after the Big Bang. Strings of particles lead to current developments in 'string theory'. Gravitation, electromagnetism and weak and strong nuclear forces hold the universe together, these forces probably resulting from an exchange of particles. The particular strength of these forces is essential to life as we know it, and any small change might make the universe sterile. This is all part of the definition of nature.

For the purposes of developing the above systems network model of technology, the important definition of nature is

The set of naturally occurring materials,
naturally occurring configurations of materials,
naturally occurring organisms,
naturally operating processes

So in our definition, nature comprises three subset classifications of:

- physical entities
- physical processes
- more elusive metaphysical abstractions

The following classification content is illustrative rather than comprehensive.

1. **Physical entities** include
   - all atoms, electrons, and subatomic particles
   - all planets
   - all natural chemical substances
   - all geographic land masses, seas and atmospheres
   - all minerals and chemicals, organic and inorganic, including petrochemicals
   - all plant and animal life
2. **Physical processes** include
   - all movement
   - climate
   - physical 'laws': for example, gravity, Newtonian mechanics

- chemical reactions
- organic life and growth
- radiation

3. **Metaphysical abstractions** include
   - energy, transformable into equivalent matter
   - time
   - force
   - field
   - propensity
   - mathematics, including infinity and the infinitesimal
   - deductive logic

This is a rather different definition of nature, more specific and relevant to the philosophy of technology we are developing, than to other more general meanings of nature. The first two categories of physical entities and processes will be familiar, but the inclusion of metaphysical abstractions may not be. Energy is metaphysical, measured in Joules, but is incorporated within a physicalist view of the world by its transformability into matter, as in Einstein's famous $e=mc^2$ equation. The Joule is equivalent to a measure of work. One Joule is one Newton of force moved through one metre, or the passing of one ampere of electric current through one ohm of resistance for one second. It therefore incorporates force which is another of our metaphysical abstractions.

3.2.1.1 Time, force and field in nature

Conceptualising time has proved elusive. Newton considered it a real measure, Kant a means of our perception, Einstein a dimension of space-time. Explicating time is beyond the scope of this book, so we simply note that nature exists in, and works to, a time dimension. Time is a characteristic of nature that does become harnessed in technology. Alongside labour and land, it is also a resource in economics.

Including force and field as metaphysical abstractions within the definition of nature may be challenged. However, whilst Isaac Newton analysed the metrics of gravitational force (proportional to the product of mass and inversely proportional to the square of distance), he did not explain the ontology of gravitational force. How does it exist, and by what means does it operate? Since we still have no answer to these questions, force remains defined as a fundamental metaphysic of nature. It certainly gets pervasively incorporated into technology. The same is true for Maxwell's electromagnetic field. Again, we can measure its metrics, laws and equations, but we do not know how it is generated or how it operates.

3.2.1.2 Probability in nature

Our definition of nature then includes metaphysical 'propensity'. This is Karl Popper's word, included in the title of his short book 'A World of Propensities'.[63] Popper was concerned that the common definitions of probability were insufficient, particularly for single event probability. He therefore coined the word 'propensity' to represent the probabilistic nature of events. Whether nature is deterministic or probabilistic (stochastic), or both, is a major question in the philosophy of physics. Prior to the Enlightenment generally, and to Newton specifically, humanity came to regard nature as sometimes regular, for example in its days and seasons; sometimes event-correlated; sometimes determined, evidenced for example by the human response of worshipping the sun; but largely unpredictable and stochastic, for example in the incidence of disease and mortality, earthquake and lightning, which were attributed to a capricious deity. Newton in physics, and his counterparts in medicine, radically altered that world view to one in which cause and effect had established a determined universe. The reliability of this new paradigm was apparent in the success of it application. Rationality reigned. Modernity was based on the paradigm and flourished.

Newtonian physics is deterministic; we are able to calculate the location of moving bodies after they have collided by solving the equations of momentum, force and friction. We can calculate the exact time and height of tides and lunar eclipses. We have built our engineered world on the basis of Newtonian deterministic physics. It is core to our life support system. How much of our reliance is based on objective deductive logic which can never fail because it is necessarily so, and how much on inductive logic which can misfire, and can take opposing conservative or radical interpretations, is unclear.

At the same time, we encounter probability in nature. The usual example is the tossing of a fair coin, which is a mechanical rather than natural exercise, but which does invoke the natural phenomenon of probability. Huge mathematical models can be built from probabilistic equations. But the core question is whether there is an irreducible stochastic element in nature? Or, if I knew everything, including all data, all phenomena, and all relationships and processes, that is, if I had a 'God perspective', would everything in nature then be deterministic, so that I could explain and forecast all physical events? For example, whilst I can forecast the tide next week, I cannot forecast the weather with the same accuracy. Is this simply because I lack a sufficiently sophisticated meteorological model, or because there is an element of irreducible probability in weather patterns? I may bump into a friend in town. The encounter is unexpected and therefore appears probabilistic, but at the same time, had I known all the causal variables in my friend's life that day, then I could presumably have forecast that he would be in exactly the same location in town, and at the same time, as me. So is probability simply lack of knowledge? Similarly, a leaf falling on my head as I walk along seems to be a stochastic event to me, but if I had known all the causal factors, of the weight of the leaf, the degree of sap reduction and dryness in its stem, the particular wind force it experienced at that moment, then presumably I could also have predicted its fall accurately.

Quantum theory challenged the deterministic Newtonian world. The location of subatomic particles after undergoing the famous two-slit experiment, appeared to be indeterminate. Particles appeared to be entangled, simultaneously in two locations, or affecting each other. Heisenberg formulated his 'uncertainty principle', and Niels Bohr developed the 'Copenhagen interpretation' of the phenomenon, representing the location of the particles as a probability distribution rather than a single determined point with definite x, y, z coordinates. Einstein, as a devotee of a Newtonian deterministic world, objected. God, he famously said, does not play dice. David Bohm in his 'Causality and Chance in Modern Physics'[64] also disputed Bohr's Copenhagen interpretation. As in some of the examples we considered above, Bohm suggested that as the scope of observation changes from the micro to an ever-wider macro level, then the apparently deterministic/stochastic mix also changes towards greater determination. If knowledge or the scope of observation is limited, then more events, for example, whether it has rained today, will appear to be stochastic, But as knowledge and the scope of observation extend, then it might be possible to model the climate or even the universe sufficiently to predict today's rain. Bohm is right in this, but it still leaves us with the same problem of whether, if we had total universal knowledge, if we were god, an irreducible stochastic element would still remain?

Other considerations point to the conclusion that there is an irreducible stochastic element in nature. Many variables in nature demonstrate a standard normal or other classical probability distribution in their values. The height of the human population is one such example. Whilst it is possible to explain why any one individual is the height they are, it is difficult, if not impossible, to explain why the population variable value follows such a specific normal probability distribution curve. The bell-shaped curve appears to be a fundamental part of nature.

This is what interested and troubled Karl Popper, who therefore, in his analysis of single event probability, developed the concept of 'propensity', proposing this propensity as a third metaphysic in nature, alongside Newton's gravitational force and Maxwell's electromagnetic field.

Interestingly the same issue arises in the philosophy of economics. Robert Skidelsky, the leading biographer of Keynes, in his 'Keynes – The Return of the Master', writes 'The centrepiece of Keynes' theory is the existence of inescapable uncertainty about the future... Classical economics was the illegitimate offspring of Newtonian physics.'[65] Keynes objected that irreducible uncertainty interrupted the 'Newtonian' equations of supply and demand, which were supposed to regulate the economy and bring everything to equilibrium, so that involuntary unemployment could result.

3.2.1.3 Mathematics in nature

Probability is therefore an important part of our definition of nature. But how does this extend to the claim that mathematics is part of nature? To justify this claim, we reverse the question by asking whether mathematics is 'objective', whether it is endogenous to humanity or exogenous, whether it is mind dependent or mind independent, whether it is invented or discovered. This is the great question of the philosophy of mathematics, but one to which the philosophy of mathematics pays insufficient attention.

Most of the philosophy of mathematics is focused on the philosophers of mathematics and is more of a history of mathematical thought, from Plato to Hilbert, Goedel, Frege, Russell, and Wittgenstein. The closest attempt to address the question of the objectivity of mathematics is by Crispin Wright and Stewart Shapiro. In his 2007 paper 'The Objectivity of Mathematics',[66] Shapiro discusses Wright's 1992 paper 'Truth and Objectivity', where Wright had proposed four criteria to test for the objectivity of mathematics.

The four criteria proposed by Wright are

i) epistemic constraint
ii) cosmological role
iii) cognitive command
iv) the Euthyro contrast

Translating these into simpler language, Wright suggests that mathematics is shown to be objective if

i) not all mathematics can be known, and/or
ii) mathematics has wide, generic, sub-conscious application, and/or
iii) there can be no disagreement about a mathematical truth (so that any disagreement must imply one party is in error) and/or
iv) any subjective judgment is involved in determining a mathematical truth

Against these criteria, Shapiro concludes

i) not all mathematics can be known since we know that an infinite number of prime numbers exists, but we cannot cognitively know all these prime numbers
ii) mathematics does have wide generic application, for example a tiler cannot lay a prime number of tiles to fill a square area, even though the tiler may know nothing about prime numbers
iii) that there can be no disagreement about mathematical truth is a reasonable null hypothesis, since where disagreement occurs, it is impossible to say whether it is due to lack of objectivity in mathematics, or an error by one party to the disagreement
iv) subjective judgment is not applied in mathematics, since for example, we do not judge whether a number is a prime number or not

Shapiro concludes that mathematics passes all Wright's tests for objectivity, but that it only needs the first criterion of epistemic constraint to be shown to be objective. Michael Dummett, however, disagrees, maintaining that all truths, including mathematical truths, are knowable. Mathematics, Shapiro concludes, is discovered and not invented. It certainly feels that way, he says, even though he admits that his view is not conclusive.

3.2.1.4 Purpose in nature

Nature appears to incorporate some coding, for example in its inclusion of objective mathematics and deductive logic. But at the same time nature appears to have no purpose. Nature is non-teleological. This claim will be disputed by readers with a religious faith, who believe in a God who acts with purpose and who wrote purpose into a created order. However, observation of the way in which nature works does not reveal purpose. Purpose would not be a leading hypothesis to interpret or explain the phenomena of nature. Attempts to distinguish purpose at work as an explanatory mechanism behind nature's phenomena usually lead to confusion rather than explanation, since there appears to be no consistency. If a belief in purpose is to be held, it has to be superimposed as an external exogenous interpretation of nature: it cannot be derived endogenously from a study of nature itself. There may be driving forces at work which create sub-purposes. For example, the urge and drive to survive leads to predatory behaviour between the species in the food chain, and this creates the subsidiary purpose of succeeding in the kill. But purpose is exactly that: a subsidiary definition and not primary, since nature – whether inanimate, or animate plant, or animal – cannot state its purpose. Life is existential. The absence of purpose in nature is highlighted by the existence of probability discussed above. If there is an irreducible stochastic element in nature, then this argues strongly against the inclusion of purpose in nature. Purpose and probability are mutually exclusive.

3.2.1.5 Infinity in nature

Nature does however incorporate infinity and infinitesimality. The cosmological universe appears to continue physically infinitely. Number systems are theorised to stretch to an infinite number of prime numbers. Time appears to continue 'forever' both backwards and forwards. We can subdivide physical entities and certainly mathematical measurements forever without being certain that we will ever reach the smallest indivisible physical particle. And if we do, why do mathematical numbers continue to be divisible beyond this potential smallest particle? Most of the philosophy of mathematics, including set theory, was developed to seek to address and express the concept of infinity.

A strange effect results when we seek to combine probability and infinity in nature. If there is a finite possibility of any event occurring, then given infinity in space-time, each and every possible event is occurring an infinite number of times in that space-time. This includes the event of my writing this book, or the reader reading it. We cannot prove that this is not true, but it does seem extremely counterintuitive. It may be the consideration which ultimately challenges and breaks the easy assumption of an irreducible stochastic element in nature combined with infinity.

In this discussion, we have seen two ways in which nature and humanity diverge despite their congruence in reality. Nature is infinite and lacks purpose, whereas physically, mentally and in life span, humanity is finite, but has a sense of purpose. As we examine the symbiosis between nature and humanity below, these two important distinctions need to be borne in mind.

3.2.1.6 The nature of nature

Nature is objective. But our knowledge of nature, and therefore our definition of it, is entirely contingent. For example, this 'knowledge' can be incorrect: a widespread view that the world was flat did not make it so, but may well have determined humanity's behaviour and its artefacts. There is a difference between nature as it really is, and nature as we conceive it to be. Experimental data from quantum mechanics appears to reveal a subatomic world which is indeterminate and stochastic, but we cannot be sure that this is really the case, or whether the limitations of our instrumentation make it appear so. Initially, nature is objective *and* independent, but as technology and the human practice it enables progresses, nature becomes co-determined with humanity. Ecological effects follow human behaviour. Landscapes are radically changed, animal species depleted by human predation, atmospheric balances altered. Humanity is part of nature, so the whole interaction of nature and humanity is endogenous. But for the purpose of the exposition of technology, and to understand the nature of the interaction between nature and humanity, we have arbitrarily defined nature and humanity as separate real entities.

What we have sought to show is that nature is not only physical. There are important metaphysical elements in nature, including time, force, field, probability/propensity, mathematics, and deductive logic. These may be shown by further scientific research to be expressed and coded in terms of the arrangement of physical elements, but the arrangements of such structuralism are additive to physicalism. Since nature itself can be shown to be metaphysical and not only physical, then it is reasonable to ask whether humanity also shares a metaphysical dimension of existence. We shall explore this issue later in the discussion of humanity as an element of the network systems model of technology, and in the section exploring the interactions between entities in the model.

Humanity encounters nature and engages with it. Initially humanity engages nature with labour. This engagement, however primitive, has to include some know-how, some technique. Labour is therefore both necessarily physical and intellectual, with the intellectual element proportionately increasing as know-how is accumulated. The raw position of humanity in nature without any mediation by science and technology which the garden of Eden myth seeks to represent, is difficult to imagine in its extreme. Perhaps human beings first existed without technology in an idyllic setting, but the harsh climate in which many humans now live requires some elemental technology for mere survival. Minimal primitive technology of spear and stone left humanity very exposed in its interaction with nature. Nature then figured in Tennyson's words 'raw in tooth and claw' and life itself in Hobbes words 'nasty, brutish and short'. Population numbers were low by today's measures, life expectancy was very short, lifestyle very limited and the standard of living mere subsistence. It is difficult or impossible to guess the human experience of that time. In barren or hostile environments the struggle to survive presumably predominated. Disease and mortality were high. In more fertile environments shelter would have been less essential and provision more available. It could be possible that life expectancy was also then higher. These are interesting questions for anthropology.

The process set out in Figure 3.1 demonstrates the dynamic of technology rather than examining any one static configuration of humanity, nature and the artefacts of science, technology, economy and society. From Ricardo on, economists have identified land, labour and time as the three fundamental economic resources subject to scarcity and therefore with a positive price in the economic system. Humanity takes one of these, its labour, and engages this with nature to produce output over time. The output can be physical or it can be metaphysical knowledge. In particular, intellectual and practical labour creates the artefact of science.

3.2.2 Science

Science is knowledge about nature. This knowledge is initially 'knowing that'. We get to know that there is a cycle of day and night, that the planets change relative position, that plants grow in a combination of rain, sun and soil, that objects fall down towards the ground. What we do, and how we do it, adjusts to this 'knowing that' knowledge. This knowledge may be known explicitly and cognitively, or it may be known implicitly, almost subconsciously, in the way that a language and its grammar are known implicitly to a native speaker, but explicitly to a foreign learner. Swapping between explicit and implicit knowledge is a complex process, as anyone who has learnt to sail or ski knows. 'Knowing that' is a documentary epistemology and is included in science. In much of the human life experience, 'knowing that' science is more familiar, and a more regular guide to our action, than 'knowing how'. 'Knowing that' science feeds into technology just as much as 'knowing how'. There are very many technologies of which the user simply does not know how they work. This is true for many of us in our use of cars and televisions. Even those of us who have some knowledge of how these devices work will find some component whose function we do not understand and cannot explain. We simply use the device. We are alienated from the technology. However, this also applies to the technologist. It is not at all necessary to know how a scientific process works in order to apply it to a technology. We might need to know that it apparently always works, that it is necessarily so. If not, then we will need to know under what exact conditions it reliably works, or what probability attaches to it working across a certain performance range.

'Knowing how' or know-how is the second level of science. But at the same time, 'knowing how' still retains its 'knowing that' status. For example, we begin by knowing that objects fall to the ground. This is sufficient 'knowledge that' to enable us to act. Then we discover from Newton that this happens because every body in the universe attracts every other body with a force that is proportional to the product of their masses and inversely proportional to the square of the distance between them. To some extent science now knows how objects fall to the ground on planet earth, and why sea tides happen with a regular timetable, and to exactly predictable heights. But science is unable to explain how this happens, and, as we have seen, is limited to saying that gravitational force is a metaphysic of nature. However far we progress in research to 'know how' a natural phenomenon works, we are able to work backwards to ask how the explanation works, and we face the well-known infinite explanatory regress that is familiar in any conversation with an enquiring child.

Nevertheless there is an important distinction between 'knowing that' and 'knowing how', and that is that 'knowing that' is specific, whilst 'knowing how' is generic and potentially transferable to other configurations of phenomena. For example, once science has shown that many naturally occurring minerals are oxides of basic metals, and can be reduced to the basic metal by reduction, and has then shown that carbon in high temperature processes, as with iron ore, or electrolysis, as with bauxite, can effect this reduction, then the 'knowing how' becomes potentially transferable and reconfigurable to other contexts. This is the power of technology, that it can create multiple or even exponential outcomes from a singular 'knowing how' of elementary science. It is why technology is more extensive than the nature on which it is, at the same time, entirely dependent.

'Knowing that' can be simply descriptive, but 'knowing how' introduces the concept of cause and effect. There is a huge body of literature in the philosophy of science as to how cause and effect is conceived.[67] It is crucial to distinguish between correlation and cause. A simple claim of causation of the type that cause C causes effect E, that is $C \rightarrow E$, is difficult to distinguish from mere correlation. We can therefore add constraints, for example of time asymmetry, such that C is always followed in time by E, that is, $C_t \rightarrow E_{t+1}$. Or we can add the condition of a differential – that is, that a change in C is always followed by a change in E, so $\partial C_t \rightarrow \partial E_{t+1}$. Alternatively, we can add a probability requirement, that there is a probability that a change in C causes a change in E, so $+/-\partial C_t \rightarrow P(+/-\partial E_{t+1})$. This is the Bayesian criterion, and suffers from the need to justify the definition of prior probabilities, and to cope with the effect of old evidence whose prior probability is 1.

In establishing claims of cause and effect in science, or rather in testing hypotheses for cause and effect, technical care has to be taken to avoid misleading traps such as 'Simpson's paradox', by which grouped data can appear to show different and even opposite correlations to the same data disaggregated. Techniques such as ensuring 'ceteris paribus', or Reichenbach screening, are important in establishing causal relationships through statistical research. Statistical regression techniques such as 'Generalised Least Squares' are often used to research cause and effect from observation data, and are the basis of well-known statistical research packages such as SPSS, originally developed for medical research of cause and effect from patient data. These techniques propose that a linear or non-linear mathematical formulation exists between the observed variables, and that if such a function is fitted to the data with the least divergence of the data from the function, then a causal relationship has been established. However, this is not certain. A very complex mathematical function could be defined which met every data point with zero total divergence, or total 'fit', but this would not be taken to represent a causal relationship. Care also has to be taken with such techniques to handle multi-collinearity where proposed independent causal variables are themselves causally related and render the estimation matrix singular. Probability causation tests whether the probability of the effect is higher with than without the proposed cause, that is, C causes E iff $P(E|C)>P(E|\text{not } C)$. Various screening and test methodologies are then applied to isolate variables in order to refine this test for probabilistic causation.

This brief review does demonstrate that claims of cause and effect are not simple to test or establish. This warns against too easy popular media claims of cause and effect being made of the sort that 70 per cent of people involved in car accidents drink coffee for breakfast. Inductive logic is often relied on to the effect that, if apparent cause and effect has been observed very many times, then it can be hypothesised in a scientific theory and technology application. Inductive logic however does not include any explanation as to how the supposed cause and its supposed effect are related. This renders inductive logic potentially unreliable. There are always conservative or radical interpretations available. A new aero-engine may have performed perfectly for thousands of test hours. The conservative interpretation is that it will perform perfectly for the next hour, but the radical interpretation is that it may fail. We rely on inductive logic every time we don't have a detailed explanation for a phenomenon, and in all such cases we are susceptible to the conservative or radical outcome. A cause and effect relationship is greatly strengthened epistemologically if we are able to suggest *how* the cause causes the effect, – to move from 'knowing that' the causal relationship appears so, to 'knowing how' it is so. 'Knowing how' is therefore a stronger epistemology than 'knowing that'. Inductive logic relies on a 'knowing that' epistemology, and is contingent – it depends on other variables and is not necessarily so – whereas deductive logic engages a 'knowing how' epistemology, and is necessarily so rather than contingent: hence its greater strength.

3.2.2.1 Science or not science – Karl Popper

Karl Popper was a leading figure in the twentieth century philosophy of science.[68] He sought to define science; to determine what is science as opposed to what is not science. For the purposes of the role of science in our systems network model of technology, the important highlights of Popper's thought are

- falsification rather than verification as the test of science
- science thereby defined against pseudoscience
- the role of inductive logic
- propensity as an interpretation of single event probability

Popper was often driven in his thought by specific events or considerations which led him to more general hypotheses. For example, the Austrian communist party's response to the police shooting its supporters, some of whom were Popper's friends, in 1918, was that such a cost was necessary for the revolution. This offended Popper, and caused him to shift from historicism, the view that there are theories of history, and specifically to criticise Marx's theory of history as pseudoscience.

Equally, Popper was impressed by Einstein's general theory of relativity, which, from Einstein's 1911 paper, was able to predict the exact gravitational bending of light in the solar eclipse of 1919, thus fulfilling the requirement of an empirical test for Einstein's theory proposed in 1915. Popper was equally impressed by the double-slit experiment of quantum mechanics, leading to Neil Bohr's Copenhagen interpretation of indeterminate positioning of the emitted particles, and this drove Popper to formulate his propensity hypothesis.

Popper identified the ability to generate falsifiable implications which could be tested empirically, as the hallmark of true science. He combined this with his distaste for the psychology of Freud and Adler to establish a definition for science which excluded Marx's theory of history as well as Freud and Adler's psychology. However, intellectual propositions which are driven by motivations other than truth seeking, are usually of dubious value, since the motivation introduces a distortion.

So Popper rejected positive verification as the test of science, specifically of scientific theories. He saw this as too weak a criterion, and insisted that science should frame its hypotheses to be capable of generating implications, which in turn should be capable of falsification by empirical test. In his 'The Logic of Scientific Discovery', Popper proposed that to qualify as science, theory has to satisfy the criteria of reproducibility, refutability, falsifiability and predictive power. It also has to respect a process of tentative hypothesis formulation and refinement, rather than making final truth claims. Hypotheses should be constantly formulated, tested for their empirical implications, and then refined.

This therefore allowed Popper to embrace Einstein but reject Marx, Adler, Freud, and astrology as science. As a result we then have a dualism of (1) science proper, which passes Popper's test, and (2) pseudoscience, which fails. The question is whether this dualism adds anything? It might in fact be argued that on the contrary, it reduces understanding by imposing false criteria, and by rejecting theories according to some meta-criterion that they are not science, rather than examining the theories themselves intrinsically, in their own right. It can too easily become a political weapon, if used according to some original purpose to exclude Marx and Freud from consideration as science. It can also become intellectually lazy to dismiss a theory simply on the claimed grounds that it is not science. It can be seen that it is more difficult for theories of history, biology and probability, particularly single event probability, to pass Popper's test to be considered science. History cannot be replicated in a laboratory, and so falsifiable theories of history, for example in economic science, would have to rely on back-forecasting ability to test as science. The same is true of evolutionary biology, which has few data points and few implications which are falsifiable within the lifetime of any research project. Single event probability cannot by definition be replicated and theories of it cannot therefore be tested empirically in the same way as Einstein's theory of general relativity was.

Indeed Popper may not have opposed theories of history, biology, and probability, and did in fact develop further theories of 'propensity' to address the issue of single event probability. He simply did not classify them as science. But this casting of such theories into the wilderness adds no value. Indeed it leaves them bereft of any rational methodology, because it suggests that they are incapable of, unsusceptible to, such rational method. How are we then to evaluate intellectual theories of history, biology, and probability? It was over the question of Darwin's theory of evolution that Popper's definition came unstuck. Popper began by accusing the theory of tautology, because the only definition of 'survival of the fittest' was that it was the fittest who survived. The theory needed an independent definition of fitness other than the fact of survival, to avoid tautology. Moreover it lacked the ability to generate falsifiable implications. Popper later 'recanted' of this critique of Darwin but his reasons for withdrawal are vague and unconvincing. The real unavoidable conclusion is that Popper's test for science vs pseudoscience is vacuous. Marx, Freud, Adler, Darwin and astrologers are all free to generate theories worthy of consideration. We can say with Popper that there are some branches of thought which can generate theories capable of empirically falsifiable implications, and these 'sciences' can follow

Popper's methodology of 'critical rationalism'. Other theories, which are not capable of conforming to this methodology, remain valid, and other methodologies for their development will have to be applied. Testability is in any case a weak claim since a theory can comprise two concepts; one reasonable, and one not, such that the reasonable element achieves testable implications for the theory and appears to endorse the unreasonable element. Falsification also has multiple interpretations: either the scientific theory can be wrong and should be amended or abandoned or, as Duhem pointed out,[69] the assumptions may be incorrect, or the instrumentation may be misleading and improperly calibrated, or indeed, the data may be incorrect, as happened when the planet Neptune was omitted from early work with Newtonian theory. Falsification is not therefore as clear and easy a criterion as Popper suggested. Popper also rejected the use of inductive logic in science, a theme we discuss below.

3.2.2.2 Science – theory or paradigm? Thomas Kuhn

Thomas Kuhn, in his 1966 'The Structure of Scientific Revolutions',[70] took a very different view of the process of science to Karl Popper. Popper adopted a 'bottom up' perspective, where individual theories are subject to his falsifiability criterion, and only ever have standing as hypotheses, not as final truth. Kuhn however took a 'top down' view, claiming that science does not develop incrementally, cumulatively, and continuously, but radically and dramatically – by 'paradigm shifts'. Within any one paradigm in the historical development of science, what Kuhn called 'normal science' is occupied with 'puzzle solving' *within* the paradigm, without questioning the paradigm itself. Popper's process is therefore 'bottom up' whilst Kuhn's is 'top down': Popper's criterion is falsifiability, Kuhn's is puzzle-solving capability. According to Kuhn it is only when anomalies in such puzzle solving accumulate that a paradigm shift occurs, and this shift is total: the whole of the paradigm is completely replaced in every detail by the new paradigm. Even terminology becomes incommensurate between old and new paradigms, so that for example, mass connotes differently in Newtonian and quantum physics. This renders the established intellectual methodology of dialectic, where thesis and antithesis formulate a synthesis, redundant and impossible, which is again a major philosophical challenge. For Kuhn, paradigms are integral and holistic. Paradigm change is an 'all-or-nothing' process. Paradigms are competitive and mutually exclusive. 'Einstein's theory can be accepted only with the recognition that Newton's was wrong', writes Kuhn. Paradigms, he claims, never reduce to special cases, one of another.

Interestingly, Popper and Kuhn come to some shared conclusions, for example on the virtue of astrology: Popper dismisses it as pseudoscience due to its inability to generate falsifiable implications (even though as Thagard was later to point out,[71] this is not necessarily true), whilst Kuhn dismisses it because it was displaced by the paradigm shift of the Copernican scientific revolution. Otherwise, Popper's and Kuhn's were competing interpretations of science: scientists were claiming that talk of 'theories' had been abandoned in favour of talk of 'paradigms'.

More importantly, Kuhn introduces sociological, historical and psychological factors into his account of science. This sounds a reasonable proposition, since science is developed in sociological and historical contexts and cannot avoid their impact, even in its experiments in a hermetically sealed class 10 clean room. Scientists are human and susceptible to holistic elements of humanity, including emotional commitments as well as rationality. They necessarily operate within human institutions which themselves are subject to well documented and analysed sociological forces.

The main question arising from this is to what extent the role of rationality is diminished or compromised in Kuhn's analytic, compared to its centrality in Popper's 'critical rationalism'. Kuhn also denied realism in science, claiming that science is instrumental and established for its capacity to solve puzzles, rather than for conformity to literal real-world reality. It is these aspects of Kuhn's theory, rather than the paradigm shift analytic, which are radical. The public awareness of science is that it is the ultimate incarnation of rationality, that it is entirely evidence based, that society conforms to science, rather than science to society, and that it models exact physical reality. This may well also be most scientists' view of science as they know and practise it. Philosophically, Kuhn's model is also radically different, since it challenges the greatest paradigm shift of all time – the Enlightenment, the Age of Reason from which rationality was held to triumph against all superstitions, feudal powers, and human emotions and institutions. Kuhn is very definitely challenging this position.

Kuhn's interpretation of science can be summarised as

- progress in science is discrete, not continuous
- science does not implement logical positivism, but is subject to social constructivism

- science does not follow the dialectic process of thesis, antithesis and synthesis, but an all-or-nothing process of incommensurate, combative, mutually exclusive paradigm
- science is not pure logic, but is determined by holistic factors and institutional processes
- specifically science is subjective and political
- rational science is constrained – paradigms rule
- science is instrumentalist, not realist
- the criterion for the acceptability of science is pragmatic – how well it can solve puzzles

Is Kuhn right? He is echoing and further substantiating similar views expressed by Max Planck, the discoverer of black body radiation, and by Einstein. Planck famously said that 'A new scientific truth does not triumph by convincing its opponents and making them see the light, but rather because its opponents eventually die, and a new generation grows up that is familiar with it'[72] and Einstein that 'It is the theory which decides what we can observe'. Both of these views appear to support Kuhn in that either science is not rationality-led, or that it is theory-led rather than starting with phenomena to formulate theory in a rational framework.

We begin by examining Kuhn's claim that dominant paradigms govern science and shift over time. My contention is that Kuhn's methodology is weak, and insufficient to justify his conclusion. He works by stating his theory of paradigm shift in science, and then illustrating it with random anecdotal examples. He admits openly that this is his methodology in chapter XI of his book entitled 'The Invisibility of Revolutions', where he writes 'I have so far tried to display revolutions by illustration, and the examples could be multiplied *ad nauseam*'.[73] He blames the publication and use of science textbooks for 'systematically disguising ... the existence and significance of scientific revolutions' and boldly states that 'only when the nature of that authority (i.e., of scientific revolutions) is recognised and analysed can one hope to make historical example fully effective'.[74] This is a strange methodology. He wants his view to be made 'authoritative', although he doesn't say how this is to be effected, so that then all examples will conform to his theory. This is theory-led in the extreme, and highly subjective, just as Kuhn claims science to be. Kuhn's examples are chosen at will, and are not a structured sample. He allows himself to choose whatever examples he feels support his point, and fails to even mention other cases. In this he deviates from Popper's methodology, since he does not test his hypothesis for falsifiability.

My core complaint is that Kuhn fails to set out any single scientific revolution in full detail and demonstrate that it follows his rules for paradigm change; for example, total change of all aspects of the preceding paradigm, incommensurate definitions between the succeeding paradigms, etc. The best that he offers are fleeting very partial references to claimed revolutions, mainly the Copernican revolution. He refers repetitively to the discredited phlogiston theory in chemistry, and hails Priestley, Lavoisier and Dalton as initiators of the revolution in the science of chemistry. But he totally fails to mention the Russian chemist Dmitry Mendeleev who developed the periodic table of the elements, and successfully predicted the existence of elements previously unknown. In this omission, which is endemic to Kuhn, he exemplifies his own claim that science can fail to observe rationality through the effect of human traits such as nationalistic prejudice, here omitting the Russian contribution to the science of chemistry. Even here in his chosen example of the history of chemistry, whilst he documents this well-known example, he does not show how this is claimed to be a paradigm change rather than an incremental discovery in the science. Perhaps it was, but Kuhn certainly does not show this. He fails to consider any other account or interpretation against which to test his view of paradigm change. The Darwinian revolution merits one page in Kuhn's book[75] as does Popper's competing view of science.[76] The major Keynesian revolution in economic science is not mentioned at all. Both in chemistry and biology, Kuhn's consideration of the data of progress of the science is insufficient for the framing or defence of his hypothesis. In his postscript, written seven years later in 1969, he does not consider that these inadequacies require any attention, writing that 'on fundamentals my viewpoint is very nearly unchanged'.[77]

Another major deficiency in Kuhn as philosopher is his failure to mention the single greatest paradigm change of all time – the Enlightenment. This seventeenth and eighteenth century intellectual revolution taken forward by a list of eminent contributors – Descartes, Spinoza, Locke, Newton, Leibniz, Vico, Voltaire, Hume, Diderot, Rousseau, La Mettrie, d'Holbach, Kant and Hegel – gets no mention except for the use of Newton as an example of a paradigm defeated by Einstein. The overwhelming replacement of superstition and feudal power by logic, the triumph of reason, leading to the successful implementation of the Enlightenment paradigm in the project of modernity stretching from medicine to engineering technology, deserves careful analysis when speaking of paradigm change. Rather than referring to the Enlightenment, Kuhn essentially challenges its core thrust by questioning the centrality of rationality in the determination of science. He might have a point, but he needs to demonstrate it with direct reference to the Enlightenment. To challenge the role of rationality it is necessary to analyse the process by which rationality previously triumphed, and this Kuhn fails to do.

Kuhn hints at a social constructivist view of science, but he offers no detailed analytic here either, although he does comment on the nature of scientific communities in his 1969 postscript.

Kuhn challenges the objectivity of science. 'This issue of paradigm choice can never be unequivocally settled by logic and experiment alone' he writes.[78] But he does not say whether he thinks that this is because logic itself is not objective, or whether logic is in fact objective but the social process of science can obscure rather than reveal this logic.

His definition of his own concept of paradigm is extremely vague, ranging from Copernican revolution, through Lavoisier's discovery of oxygen combustion, to the discovery of X rays. Margaret Masterman[79] and Ernan McMullin[80] respectively take up these points. Having noted that Kuhn had become so much more popular than Popper that 'paradigm' had replaced 'hypothesis' in scientific jargon, Masterman distinguishes 21 different uses of 'paradigm' in Kuhn which she categorises into three paradigm types of metaparadigm, sociological paradigm, and artefact paradigm. Kuhn responds to this in his postscript by admitting some 'stylistic inconsistencies' and redefining a 'disciplinary matrix' for scientists working within what Masterman may have called a metaparadigm. Masterman's critique is essentially pro-Kuhnian.

Imre Lakatos[81] however accuses Kuhn of relativism which challenges the objectivity of science. According to Lakatos we can never be sure that a theory will not find further empirical refutation, so no theory is provable, neither can evidence increase the probability that a theory is true. One piece of evidence does not refute a theory: it either refines the theory, as in Clairault's correction of Newton's calculations to align them with the observed lunar orbit, or finds necessary changes in the underlying data; for example, the solution of Uranus's apparently incorrect orbit by the discovery of the planet Neptune. As Dudley Shapere shows in his 1971 review of Kuhn and Lakatos' critique,[82] Kuhn responds by introducing 'a paradigm independent world (nature) which presents problems that a paradigm must solve'. 'That' as Kuhn writes 'is not a relativist's position and it displays the sense in which I am a convinced believer in scientific progress'. Shapere considers that in this Kuhn has substantially modified his original position.

Ernan McMullin questions the equivalence of Kuhn's various exemplar scientific revolutions. The discovery of X rays was a minor step forward and hardly qualifies as a revolution, although Kuhn claims it to be so because of radical changes in instrumentation. The replacement of phlogiston theory by oxygen theory raised no epistemic questions in the way that the Copernican revolution definitely did, so that only the latter is rightly termed a revolution in the normal semantic sense of this word because of this. McMullin also criticises Kuhn's discrete notion of normal puzzle-solving science and scientific revolution, saying of examples in physics and paleontology that 'what we have here, I suspect, is a spectrum of different levels of intractability, not just a sharp dichotomy between revolutions and puzzle solutions'.[83] This more refined view of science appears closer to the reality of the process than the two Kuhnian processes.

As McMullin says 'Kuhn rests his case then, both for the rationality of science and for its distinctiveness as a human activity, mainly on the values governing theory choice in science. But he does not chronicle their history, disentangle them from one another except in a cursory way, or enquire in any detail into how and why they have changed in the ways they have'.[84] This last critique resonates with the opening critique of this section, that Kuhn freely makes claims without any detailed analysis. McMullin therefore criticises Kuhn's claimed epistemic values. Kuhn has set these values out as

- accuracy
- consistency
- scope
- simplicity
- fruitfulness

but denies that these determine scientific outcomes, and regards them as entirely subjective without any link to objective truth. McMullin regards these epistemic values as desirables and as means to the two great ends of predictive accuracy (= empirical accuracy) and explanatory power. He questions whether all scientific paradigms share Kuhn's five epistemic values, whether these values are totally subjective without rational justification, and whether Kuhn is correct to deny realism in favour of instrumentalism in science.

Kuhn appears to have selected epistemic values that cannot be demonstrated to lead to empirical adequacy or truth-realism. The real world for example might be inconsistent, especially in behavioural sciences like economics. Specific theories might be more accurate than general theories of wider scope, so that Kuhn's epistemic values can be mutually contradictory, and themselves fail his test of consistency. Only by induction can these epistemic values be construed to support a theory, (good theories in the past tended to embrace or exhibit these values although we may not know why), and McMullin and Kuhn, like Hume and Popper, are all uncertain about the reliability of such inductive logic.

McMullin is a scientific realist and challenges Kuhn's instrumental view of science which is judged only by its ability to solve puzzles. McMullin shows that, contrary to Kuhn's view, Copernican theory explains far more than Ptolemaic theory, so that explanatory power, unlike predictive power, is an indicator of objective truth and should be regarded as a criterion in favour of a theory. We can follow McMullin in embracing explanatory power without conceding full scientific realism, a theme we shall examine more thoroughly below. Newtonian theory has explanatory power, even though its concept of force does not refer, – is instrumental rather than real, and like Maxwell's electromagnetic field and Popper's propensity, remains at least for now a fundamental metaphysical entity (which is why Bishop Berkeley objected to Newtonian force as 'occult' and why it is correct to say that Newton formularised gravity and showed its parametric dependences without being able to explain gravitational force). McMullin also questions Kuhn's distinction between normal and revolutionary science, proposing from examples in paleontology, physics and cosmology that any distinction is only in scale, not in type.

Laudan[85] points out that scientists can validly pursue different objectives in their work. In some cases empirical adequacy is the goal, in others explanatory power. Similarly paradigms are not a holistic unity but can be developed in part. Kuhn's hermeneutic is therefore too tight and too global in insisting in singularity of objectives and theory-holism in science. McMullin and Laudan both conclude that Kuhn is incorrect in subjugating the role of rationality in science in favour of non-rational factors. Curd and Cover summarise 'The evaluation of Kuhn's work by McMullin and Laudan is largely negative. Both authors agree that Kuhn has failed to establish his more sweeping claims about the irrationality of scientific revolutions. Neither the history of science, nor Kuhn's philosophical arguments show that scientific revolutions cannot be resolved by rational argument based on evidence and shared rules. By treating paradigms as indivisible wholes and by failing to appreciate the ways in which rules and aims can be rationally debated, Kuhn has seriously underestimated the role of reason in paradigm debates.'[86]

What Kuhn might have explored but did not is the set of factors which potentially impede and qualify pure rationality determining science. These might include

- the necessary holistic nature of humanity including its scientists and academics
- the necessary institutionalisation of humanity and its science
- the almost necessary holistic nature of science with important interference effects into, or implications for, other areas of human life and thought

Scientists, like all academics, will have emotional as well as rational commitments. They almost always work within institutions whose processes may not all be only rational. And science can have interference effects, for example in the case of Darwinian evolution's impact on theology and religion in society, and these can have feedback effects which threaten the rule of pure rationality in science.

However, from the perspective of our systems network model of technology, Kuhn's view of science as socially constructed can only have limited explanatory power. Although social factors may affect science in ways to be discussed more fully later, more importantly, science often feeds into technological applications, and these have to work in the real world. This is at least true of scientific knowledge, of the 'knowing that' aspect of scientific epistemology, even if it is not necessarily true of scientific theory. Where scientific theory is itself implemented in real-world technologies, then science has demonstrated its truth in realism. If science is defined by other factors than rationality, then science may be in great moral danger, since the alternative determinants are ones of social power, potentially making science the tool of perverse regimes.

3.2.2.3 Philosophy of science

So far we have examined in some detail the definitions of science proposed by Popper and Kuhn:

- that science is a ***rational process*** of formulation and refinement of hypotheses which have to be capable of empirical implication, test, and falsification (Popper)
- that science is a ***pragmatic process*** of shifting through best puzzle-solving paradigms (Kuhn)

Philosophers of science have further concerns about the nature and methodology of science. These include

- **Induction**: whether science can validly use inductive logic or only deductive logic
- **Reduction**: whether and how general scientific theories can reduce to specific case theories
- **Realism**: whether science is or can be instrumentalist, or has to be real and 'true'

3.2.2.4 Induction

Popper followed David Hume in rejecting the inductive method for the rational intellectual method and therefore for science. Induction is based on observation. The sun has risen every morning, therefore the sun rises in the morning. We may not know how this happens, just that it does happen. There is no explanatory logic but it remains logical as inductive logic to rely on an apparently unchangeable natural phenomenon. Since it happens with observed regularity, it's logical in one sense, the inductive sense, to assume it will continue to happen, ceteris paribus – all other things remaining equal. Hume objected to the circularity of inductive logic, that is that we can only rely on inductive logic because of its apparent past success, an argument which is itself inductive. Critics of inductive logic are also correct to point out that induction has two arbitrary and opposing conclusions, the conservative and the radical interpretation. The conservative interpretation is that a regular phenomenon will continue (without knowing how it continues), whilst the radical interpretation is that this regularity will fail and the phenomenon will not occur next time, or will change in some way. This is labelled 'underdetermination'.

Curd and Cover[87] eloquently point out however that Hume, Popper, and Kuhn's scepticism about inductive logic is misplaced, in that scepticism about deductive logic is also unanswerable to a critic of the deductive method, since deductive logic itself is the only way to try to convince the critic. So all logic is ultimately incapable of independent justification by a method outside itself. Just as we don't know how conundrums which appear to defy logic exist, so we don't know why logic itself, whether inductive or deductive works. We just observe that it does, and we base our scientific theories and our technology applications on a combination of inductive and deductive logic. Indeed, the technological test for science, that it works in practice, is usually omitted by philosophers of science, who rarely look to the phenomena of practical implementations, of which they appear blissfully unaware. Plato appears to have been deluded in his view that philosophers should run society, particularly those with the specialist mathematics training which he insisted on. Philosophers like Reichenbach and Salmon[88] are exceptions to this and do accept that continuing pragmatic reliability does justify inductive logic.

Inductive logic is similar to our earlier definition of 'knowing that' science, and deductive logic parallels our definition of 'knowing how' science. It is true that our ultimate uncertainty of what we know by induction, its 'underdetermination', requires a probability qualification around inductive statements. Popper's scepticism towards induction should therefore be answered by his development of 'propensity' examined earlier. He cannot have propensity and an insistence on the hegemony of deduction. Popper later modified his position on falsifiability to allow that a theory which had withstood severe testing could be regarded as 'corroborated' though not confirmed as truth. Salmon[89] correctly points out that this leaves the scientist unable to decide to apply a scientific theory to a future context without relying on inductive logic that the theory was correct in past contexts and so will be correct in this future one.

We can therefore summarise this and our earlier critique of Karl Popper's falsification test for science in the following points

- It does not answer the 'so what?' question – declaring a theory unscientific doesn't really say anything
- It is too dichotomous: theories can range along a continuous spectrum of testability
- Inductive logic is no more difficult to justify than deductive logic
- We have no choice other than to live a large part of our lives on the basis of inductive logic

Carl Hempel's 'deductive-nomological' model for science, insisted on the inclusion of a 'law' in a scientific explanation.[90] This explanation should then be capable of prediction, and to Hempel, explanation and prediction co-exist symmetrically. Each explanation drives a prediction and each prediction incorporates an explanation. Initially in 1948, Hempel defined his scheme in terms of hypothesis and deductive logic, but later in 1962 included probability of prediction in his 'inductive-statistical' model. Hempel later modified his view to say that whilst all explanations can be restated as adequate predictions, not all predictions generate adequate reliable explanations. Peter Railton added Popper's 'propensity' to allow the inclusion of improbable events in Hempel's scheme,[91] whilst Wesley Salmon defined the 'ontic' conception of explanation to require a statement of the underlying causal mechanisms of the explanation. This approaches the 'knowing-how' epistemology of our initial definition of scientific knowledge. 'Knowing-how' remains a stronger epistemology than 'knowing-that' since it incorporates explanation and is capable of more reliable prediction, being more similar to deductive logic than to inductive logic.

Hempel therefore needs to define what is meant by 'law' for a scientific law to be incorporated in his proposed methodology. In turn, scientific law is defined in the literature either as 'regular' – which refers to observation of regular phenomena and corresponds to our 'knowing that' epistemology, or as 'necessitarian', which refers to a law including an explanation of the phenomenon, making the phenomenon and any law about it necessary rather than contingent, and corresponding to our 'knowing how' epistemology. Critics point out that regular law fails to cover laws where there is no instance, and accidental phenomena which are not due to any regularity. Dretske frames his neccessitarian laws in terms of relationships between properties which he regards as universals.[92] Nancy Cartwright thinks that scientific laws don't describe how bodies actually behave, and in this sense are not 'true'.[93]

James Ladyman[94] offers a neat summary of the combinations of inductive logic and rationality in scientific thinking by author as

Table 3.1 A typology of rationality in science

	Inductive	Non inductive
Rational	***Carnap***	***Popper***
Non rational	***Hume***	***Kuhn***

3.2.2.5 Reduction

Reduction is the possibility that general scientific theories reduce to more specific theories as some parameter approaches zero. So, for example, quantum mechanics may reduce to Newtonian mechanics when the ratio of Planck's constant to system size asymptotically approaches zero: for large systems, Newtonian mechanics apply as we know they do. There are similar questions as to whether classical genetics reduces to modern molecular biology, whether Newtonian mechanics is a special case of Einstein's theory of relativity, whether 'special sciences' like chemistry reduce to physics. The main point of discussion is whether one theory reduces to another, or whether one theory replaces another. Thomas Kuhn denied this, claiming that scientific paradigms are holistic and entirely mutually exclusive.

However, Michael Berry, Professor of Physics at Bristol University, demonstrates that the mathematics of different theories addressing the same physical phenomena, predict the asymmetry of theory which we see in the real world.[95] This is due to mathematical singularities which are factors whose value makes the mathematics of a theory indeterminate at some point. These singularities determine the asymptotic limit of a theory. This means roughly that some mathematical term in one theory asymptotically approaches zero with some specific variable, hence rendering the term insignificant and amending the theory, apparently transforming it into a rather different theory. So Berry shows how quantum theory reduces to classical Newtonian mechanics as the ratio of

Planck's constant to system size reduces towards zero. Similarly, statistical mechanics reduces to thermodynamics as the number of particles tends to infinity, or its reciprocal tends to zero. Viscous flow theory reaches a singular limit as turbulence sets in as the reciprocal of the Reynold's number approaches zero. Wave optics switches to ray optics for very small wavelengths, special relativity to Newtonian mechanics for bodies at low speed, and general relativity to special relativity. These singular limits are, says Berry, 'a general feature of physical science'. They therefore require no specific philosophical interpretation, since the link between two related theories is a straightforward mathematical necessity, and can be set out mathematically as Berry shows. The main philosophical significance is in theory application: Berry's example is that an engineer designing a bridge does not need to calculate how the constituent atoms in the structure are arranged, but can proceed reliably using standard high-level mechanical engineering theory and formulae.

3.2.2.6 Realism

To most people, science seems to be real and true. Science addresses the reality of natural phenomena and generates theories about relationships between these phenomena which are true. They are real and true because they can be repeated in experiment and practice. So the claim that science is not necessarily either real or true comes as a surprise. The debate hinges on what entities are observable or unobservable. How can we be sure that unobservable entities really exist, or do we even need to ask this question? If a scientific theory works in repeated experiment and practice, is the truth of every entity it includes important? Should every entity in a scientific theory 'refer' to a real corresponding entity in the natural world? The answer will depend on whether we adopt a pragmatic view of science, as Thomas Kuhn did – in which case the reality of all assumptions is unnecessary and an instrumental view of science will suffice; or whether we are interested in the ontological principle of science, so that all entities must be real and all theories true. We can see that the requirement of truth conflicts with Karl Popper's view that science never achieves truth statements anyway, but only hypotheses for further empirical testing. Ultimately we can question even whether what we see is true, or only the image effect of an entity on our instruments or on our eyes.

Bas van Fraasen[96] argues for 'empirical adequacy', that the realism or truth of science is not important as long as it is empirically adequate, that is it works in repeated experiment and application. The counter-argument is that, if scientific theories are not true, then their success is miraculous. But van Fraasen has more Darwinian explanations for the success of scientific theory which does not require it to be true or real.

Science can proceed by inference. For example, early atomic theory was able to infer the existence of an electron, without an electron being directly observable. More significantly, the Newtonian theory of force cannot pass the realism test, since force does not refer to any observable entity. Force cannot be observed: only its effects can be observed. Similarly with Maxwell's electromagnetic field, which again cannot be observed. So here we have two well-known scientific theories with elements that do not refer to a real counterpart. Gravitational force and electromagnetic field are metaphysical. Theories about force and fields can therefore only be instrumental science. Karl Popper claimed the same for his concept of 'propensity' which he proposed as a third natural metaphysical alongside force and field. Science therefore need not, and in some cases cannot, insist on realism in all its entities. Equally therefore, realism cannot be a criterion for the definition or practice of science.

We may conclude in all this that philosophy of science suffers from a basic incompatibility between philosophy and science. Philosophy is concerned with detailed rigorous intellectual conceptual justification, even if at times this can seem far removed from the real world or any practical concern. Science is concerned to 'know-that' and to 'know-why' and feeds this into practical technology. Philosophy is therefore demanding of science a particular philosophical rigour which science never set for itself. Science might equally examine philosophy and find it lacking, incapable of generating falsifiable implications, or of solving any puzzles. So science might judge that philosophy is not science, whilst philosophy might judge that science is not philosophy. The two disciplines might have little to say to each other, little common ground.

The Nobel Prize physicist Richard Feynman is credited with saying that 'philosophy of science is about as useful to scientists as ornithology is to birds' but as Alan Sokal has responded '... you know the famous quote from Feynman which says 'philosophy of science is about as useful for scientists as ornithology is for birds'. Most people would see that as a denigration of the philosophy of science, but I don't see it that way at all. Ornithology is not intended to be useful for birds. In principle ornithologists might, by studying the physics of how birds fly, come up with some suggestions to birds about how they could fly more efficiently, except that natural selection has probably beaten them to it anyway. In the same way, philosophy of science could come up with suggestions for working scientists, but that's not necessarily its major goal. I like that Feynman quote precisely because it's not, in my view, pejorative towards philosophers of science. It's saying that the philosophy of science is different. It clarifies what scientists do whether or not it helps scientists.'

Feynman is usually quoted to show that there is substantial cynicism amongst scientists about the value of philosophy of science. Sokal's response is correct but goes on to leave no definitional or value challenge for the philosophy of science, apart from clarifying what scientists do. He doesn't allow scientists even to comment on the value of such clarification.

So what does the philosophy of science claim to offer? James Ladyman, a leading practitioner of the philosophy of science, states in his book 'Understanding Philosophy of Science' that the philosophy of science is all about epistemology[97] – all about what science can claim to know, although he later wonders whether 'epistemological scruples' are sufficiently substantial to define a philosophy of science.[98] In-depth argument about whether and on what basis science can claim to know anything becomes rapidly arid and trivial. There are greater philosophical issues in science, for example, whether Newtonian mechanics is correct and the physical world is deterministic, or whether quantum physics has shown that there is an irreducible stochastic element in nature.

3.2.3 Technology

Technology is entirely contingent on nature, science, and humanity. It derives from ***nature*** in that there is no technology that does not consist of natural materials and processes. It derives from ***science*** which is the know how, whether primitive 'knowing that' or sophisticated 'knowing how', necessary to specify reconfigurations of natural materials and processes. And it derives from ***humanity***, since

i) human curiosity, endeavour and intellectual labour are necessary to elucidate science from nature,

ii) human intentionality is needed to activate any technology either in its development or its application, and
iii) human creativity is necessary to achieve the reconfiguration of natural materials and processes that is the very definition of technology.

Science and technology mutually interact, the classic examples being the microscope and telescope which derive from optical and mechanical science, but then enable science to observe natural phenomena invisible to the naked eye. In science, humanity has therefore already become a techno-human, demonstrating how complex the interactive definitions in this systems networked philosophy of technology are.

Whilst nature, science and humanity are inter-dependent, it is undoubted that technology has become a huge and powerful artefact which has given it the appearance of an autonomy it does not necessarily possess. The 'complex consolidation', between technologies and within technology as a whole, noted on Figure 3.1 is in fact amazing. A complex web of technologies is embodied in almost every product we use. Seismological survey, mineral mining and refining, plastics extrusion, computer design, software capability, integrated circuitry, machine tooling, precision casting, complex logistics are incorporated into almost every everyday item. Rather than technology enframing humanity as Heidegger claims, it looks rather as though humanity has enframed technology, although in reality the two are in powerful symbiosis.

W Brian Arthur in his 'The Nature of Technology', centralises this inter-technology and intra-technology combinatorial dynamic in his thesis. Rather than any one technology driving change, it was in fact the *combination* of technologies that proved to be the core dynamic.[99] Following Joseph Schumpeter, Arthur defines this as 'combinatorial evolution', the 'autopoietic' process whereby 'technology creates itself out of itself' working through 'the constant capture of new phenomena'.

He defines technology itself variously as

- 'a means to fulfil a human purpose',
- 'an assemblage of practices and components',
- 'a collection of devices and engineering practices available to a culture'
- 'a set of phenomena captured and put to use'
- 'a programming of phenomena to our purpose'

Technology for Arthur is essentially 'executable'. Technology is built as a hierarchical structure, with many lower level modules common to several higher technologies. Another Schumpeter insight is that technologies are 'clustered' into 'domains' and engineered into higher level technologies. The economy is 'constructed from its technologies'[100] and this technology/economy is never in stasis, but in a process of constant dynamic change.

3.2.3.1 A technology narrative

Vaclav Smil in his two books 'Creating the Twentieth Century: Technical Innovations of 1867–1914 and Their Lasting Impact' and 'Transforming the Twentieth Century: Technical Innovations and Their Consequences',[101] documents the extensive and intensive deployment of technology into the human world. His work is largely descriptive narrative rather than analytical philosophy. He calls the 1880s 'the most inventive decade in history' including the development of electric powerplants, lights, motors, trains and transformers, steam turbines, gramophones, cameras, the internal combustion engine to power cars and motorcycles, aluminium, crude oil tankers, pneumatic tyres, and pre-stressed concrete. In 1909 Fritz Haber achieved catalytic synthesis of ammonia, and implemented its production as the Haber-Bosch process together with Carl Bosch of BASF to produce nitrates both for munitions and for agricultural fertiliser. Smil claims that without this fertiliser, 'some 40 per cent of today's humanity would not be alive'.[102] He charts the detailed technological story of four sectors of the economy: energy, materials, production, and transportation.

The incarnation of science into technology is well exemplified in Smil's account of the development of nuclear energy. According to Smil,[103] in 1931 at the Cambridge Rutherford laboratory, John Cockroft and Ernest Walton achieved nuclear fission by accelerating hydrogen atoms into a lithium nucleus to emit two helium particles. From this, James Chadwick (in his own words) 'supposed' the existence, formation, and emission of the neutron. Rutherford, however, was sceptical of its prospect for industrial implementation and in a 1933 lecture said 'anyone looking for a source of power in the transformation of the atoms was talking moonshine'. But Leo Szilard, a Hungarian physicist who had studied under Einstein, conceived the idea that an element could be split by a neutron and emit two neutrons in a nuclear chain reaction, and suggested beryllium, uranium and thorium as possible candidates. In 1938, Otto Hahn and Fritz Strassman irradiated uranium to produce several new isotopes including barium, a process which Otto Frisch interpreted as nuclear fission. Lise Meitner and Otto Frisch published their interpretation in Nature in 1939 that 'it seems therefore possible that the uranium nucleus has only small stability of form, and may, after neutron capture, divide itself into two nuclei of roughly equal size. These two nuclei will repel each other and should gain a total kinetic energy of c 200 Mev.' This science led to the world's first nuclear reactor being commissioned at the University of Chicago in 1942 in the Manhattan project's search for a nuclear bomb. Meanwhile in 1944, Enrico Fermi and Leo Szilard filed their patent for a nuclear reactor and in 1956, the UK was the first country to commission its Magnox nuclear powerplant at Calder Hall in Cumbria which operated until its decommissioning in 2003.

Core machine technologies deployed in the twentieth century were the internal combustion engine, the electric motor, and the steam turbine – all having their roots in 1880s technology. To these were added the gas turbine, developed in the twentieth century and achieving extensive implementation in (1) powering aircraft for air transport, (2) generating electricity, with the advantages of fast response times for peak demand generation requirement, low capital cost, high thermal efficiency, and lower atmospheric emissions, (3) pumping gas and liquids along pipelines.

Materials technology produced steel, aluminium, plastics and polymers, and semiconductors. Mass automated production raised productivity immensely.

These achievements result from the network of underlying enabling technologies. So technology which is itself a reconfiguration of naturally occurring materials and processes, yields technology modules which can themselves then be further configured into the range of machinery described by Smil.

To try to understand the map of these fundamental technologies, Figure 3.2 (see overleaf) sets out the complex network of multiple technologies that feed into the human-built world of infrastructure, and the goods and services of mass consumption and enhanced lifestyle. Within each of these technology types, further layers of complexity are incorporated. For example, **mining technology** itself is sophisticated and ever advancing in its capability. Opencast mining may seem a relatively simple operation of shovelling coal or minerals from shallow deposits, but the advanced technology of the draglines used, the calculations of overburden removal, the geological analysis required, are all complex. Underground longwall mining, with hydraulic roof supports, armoured face conveyors and shearers, became ever more sophisticated, with retreat practices, automatic shearer sensitive longwall face advance, and roof bolting technology. This latter replacement of steel arches with simple bolts through the rock roof above the mineral deposit was strongly resisted by workers' unions, which accused mine owners and managers of compromising worker safety in the interests of cost reduction. So far this account would be grist to the mill of the social constructivist theory of technology. However, geological science shows that the dendritic plates are arranged in horizontal layers so that preventing them slipping horizontally by simply bolting through the layers of dendritic plate, is the important engineering requirement against roof collapse. Science in fact determined the technology outcome and not social constructivism. Underground coal gasification (UCG) is an advanced technology for converting coal into gas in situ and piping the gas to the surface. All these demonstrate how complex the technology matrix, from which humanity as the consumer is largely alienated, has become.

Refining technologies extract metals from their ores by reduction, such as iron from iron ore, or by electrolysis, such as aluminium from bauxite, and petroleum products from crude oil by hydrocarbon cracking. Whilst oil and gas are huge feedstocks to a wide range of petrochemical technologies and to the whole plastics industry, electricity is a major determinant of technological production, energy consumption, and social lifestyle.

The **power generation technologies** deployed are coal or gas fossil fuel, with coal generating twice the CO_2 emissions of natural gas. Nuclear, wind and solar are emission free, but environmental lobby groups have been split on their deployment. Nuclear power has been denigrated by political lobby groups because of the cost and danger of dealing with its spent fuel, the danger of a nuclear accident following Chernobyl, Three Mile Island and Fukushima, and its proximity to nuclear weapons technology. Wind-power creates visual environmental disadvantage, and for this reason lobbying has been as much against as in favour. Zero-emission electric power is both limited quantitatively and expensive per unit to produce, and public debate has not been able to arrive at a consistent position of electricity consumption levels, the

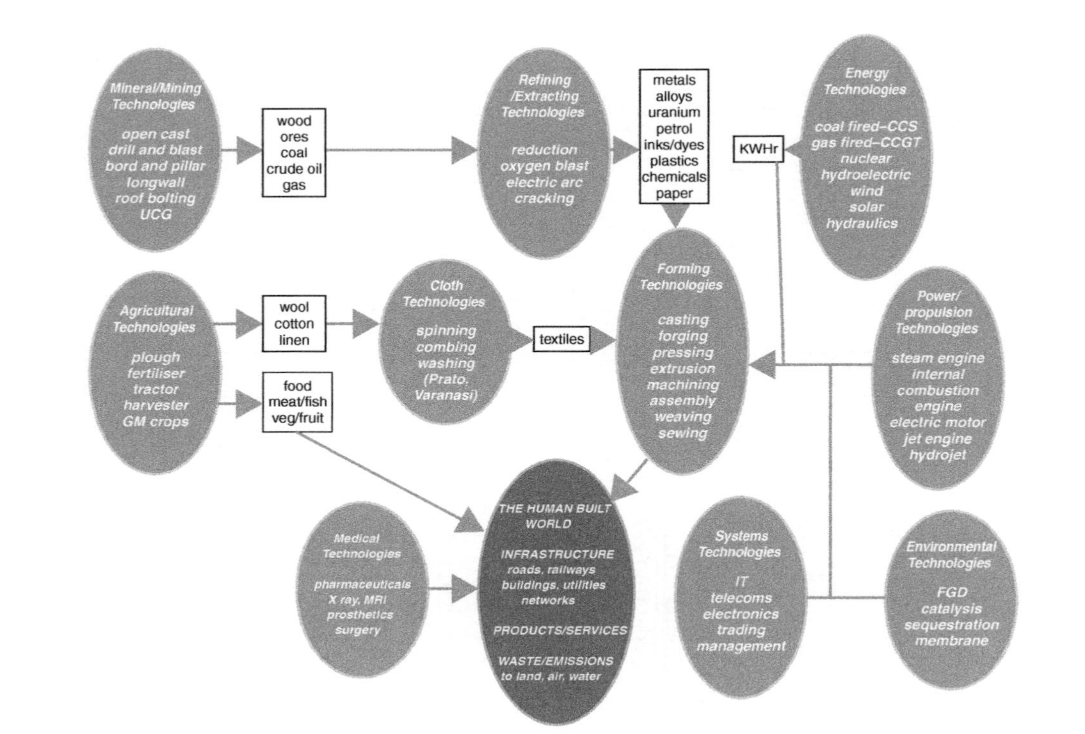

Figure 3.2 An illustrative complex network typology of technology

associated means of production, the level of emissions implied, and the consumer cost of a unit of electricity.

Meanwhile technology advances, firstly with vastly higher thermal efficiencies for fossil fuel powerplant technology which, using high temperature alloy boilers, can reach over 50 per cent for coal-fired plants and 60 per cent for gas-fired plants. Compared to previous generation thermal efficiencies of around 30–35 per cent for coal-fired plants, the CO_2 emissions of these high thermal efficiency plants are greatly reduced per unit of electricity output. These technologies are socially constructed in that society wishes low emission power generation, but they rely on science for their genesis. Flue gas desulphurisation (FGD) has dealt with SOx emissions and selective catalytic reduction (SCR) with NOx emissions. It is difficult to see how these technology shifts can be socially determined. Meanwhile gas-fired powerplants have harnessed combined cycle gas turbine (CCGT) technology where a secondary steam turbine raises thermal efficiency to 60 per cent and greatly reduces emissions. Carbon capture and sequestration technology (CCS) is now being piloted in plants such as the one at Monash, Australia. The coal industry is envisioned in that it dealt with SOx and NOx emissions effectively, and so will deal with CO_2 also. Let's hope so.

Forming technologies are where mechanisation took root. Casting technology has moved from gravity sand mould casting, through lost wax technologies to thixotropic casting. The latter technology for casting aluminium was discovered in 1926 by a Czech scientist who found that, if molten aluminium was stirred whilst cooling, then the dendritic crystal structure was broken up and a globular structure formed. This thixotropic ingot melts into a paste at a lower temperature than standard ingots, and when pushed into a die in a casting operation, does not produce porosity, meaning both that it can be machined smoothly, and also that it can be cast to near net shape, thereby reducing final machining costs. Pechiney, the former French aluminium producer researched the business case for this aluminium technology and invested in its production for carburettors, computer disk drives and aerospace components, due to its greater competitive price/performance in these applications over other casting technologies. The technology was thus economically, rather than socially or politically, contingent.

The technology for forming metal tubes – so extensively used in hydraulics and pipeline applications – developed new methods. Tubes which were made from sheet metal welded either longitudinally down the seam, or spirally to enable welding of high diameter pipeline tube, suffered from an inconsistent internal diameter surface finish. They were challenged by seamless tube formed by drawing the internal diameter over a mandril, thus achieving much higher precision of the internal diameter surface, and increasing the performance of such tube in hydraulics applications, where hydraulic seals could now operate against a highly consistent surface.

Machining technology evolved from the hand tool, where philosophies of human extension may be appropriate, to machine tools, which incarnate the concepts and capabilities of someone other than their user and arguably deskill the worker-user, to numerically controlled machine tools, and finally to advanced computer-aided design, computer aided manufacture (CADCAM) systems, which require little hands-on operational labour but extensive skilled labour in programming.

3.2.3.2 Textiles technologies

Great Britain was the sponsor of new high-productivity textile technologies, invented in the second half of the eighteenth century, and deployed through the first half of the nineteenth – both in the UK and in export markets. John Kay's flying shuttle, developed in 1733, gained widespread use and has been described as the first technology to achieve a quantum increase in productivity. This was followed in 1764 by James Hargreaves' spinning jenny for spinning yarn for the weft in weaving, and in 1769 by Richard Arkwright's water frame installed in his Cromford mill, which was also capable of spinning the warp. There were bitter patent disputes over these inventions. Kay migrated to France in 1747 to seek a better patent income there, although with little success. Thomas High had an equal claim to the development of Arkwright's water frame. Samuel Crompton's mule appeared in 1779 but it was not until 1830 that Richard Roberts' power loom was developed. Karl Marx remarked that Robert's loom would 'open up a completely new epoch in the capitalist system', but in fact the loom took a long time to become widely used, and it was 50 years before it accounted for the majority of UK cotton output. Britain was already better at invention than at innovation, at technology development than at technology deployment.

These engineering technologies led to huge output increases in UK textiles production.[104] From 1770 to 1800, UK textiles output increased by 33 per cent, but from 1800 to 1842, output multiplied by a factor of six. This resulted from the deployment of looms, which increased from 2,400 in 1803 to 100,000 30 years later in 1833. By 1841, 115 textile machinery firms in Lancashire employed over 17,000 people, and in the early twentieth century, the textile machinery sector was the largest single branch of engineering in the UK, employing 40,000 staff, three-quarters of them in six large Lancashire cotton machinery companies. After the repeal of the ban on machinery exports in 1843, the technology was widely exported. From 1860 to 1875, Platt Brothers of Oldham UK exported some 50 per cent of their cotton machinery output; in 1873, 36 per cent of this was to Russia, 12 per cent to India, and 9.5 per cent to Germany.

Kristine Bruland has documented how British textile technology export, both of machinery and skilled labour, enabled a substantial textile production industry to become established in Norway. She makes the important point that technology is often disseminated by being incorporated in capital equipment, but also that this needs matching export of skilled operators. She quotes D Jeremy's description of the result of failure to match equipment with operator skill: 'The crucial importance of the manager and machine builder in the Arkwright system and of the operative in the Crompton system ... could not be surmounted by importing machines without men. At Philadelphia, a disassembled spinning mule confounded interested parties for four years, and was eventually shipped back to Britain in 1787, leaving Philadelphians none the wiser but angrier'.[105] Capital equipment alone does not guarantee productivity: sophisticated packaging machinery installed in current Russian factories achieve only a fraction of the productivity they are capable of, due to low efficiency labour practices.

The industry has since evolved, shifted its geographic centre determined by competitive positioning, and concentrated in Italy and Korea. Textile technologies for higher quality washing and combing of wool cloth have been implemented in a small-scale artisan industry centred on Prato, Italy. Meanwhile, in a country which adds layers of technology over time without replacing older technologies, French Jacquard punch-hole card-controlled looms for weaving silk are installed in thousands of artisan weavers' homes in Varanasi, India. The textiles industry is thus fragmented in parts of its value chain, as well as highly concentrated in others.

3.2.3.3 Agricultural technologies

Agricultural technology, through its effect on productivity in food production, has been crucial to the survival of humanity, and to the development of advanced social structures. Sufficient food is essential to survival. Food surpluses drive the formation of social structures from hunter-gather and pastoralist societies to feudalism, as set out in the section on society below. An analysis of agricultural technology and agricultural productivity is therefore central to understanding the human condition, and to managing its development.

Productivity gains in agriculture through the deployment of technology have been huge. Strangely, they are largely unnoticed by the general population, who are simultaneously totally dependent on and unaware of these quantum leaps in agricultural productivity. J L Anderson in his 2009 study 'Industrialising the Corn Belt – Agriculture, Technology and Environment'[106] traces the deployment of high-productivity technology in the corn farms of Iowa. In 1945, US farmers produced sufficient output to feed themselves and 11 other people, but by 1970 were producing to feed 42 other people.

From just over 20 million acres, in 1945 Iowa farmers produced 462 million bushels of corn and 11 million pigs. By 1970, output from the same land had grown to 858 million bushels of corn and 18 million pigs. This was achieved by switching to hybrid seed, using pesticides, fertiliser, a range of new machinery, feed supplements, and optimal farming practice.

The chemical DDT was widely used to eliminate the corn borer insect which destroyed 4 per cent of the Iowa corn crop. Sprayers grew from 5,000 in 1947 to 42,000 in 1950 until borer resistant hybrid corn was introduced in the 1960s, and DDT was banned in 1970. Over 90 per cent of cattle farmers used fly spray on their dairy cows and barns. Soil insecticides were used to control rootworm. In 1949, 41 per cent of Iowa's corn farmers used pesticide, until the 1962 publication of Rachel Carson's 'Silent Spring' led to the Iowa Pesticide Act of 1964, regulating pesticides which were appearing in water tables. The chemical 2,4-D, a synthetic hormone growth regulator, was extensively used for weed control, although it had little effect on grassy weeds such as foxtail. Even aircraft spraying was used which was cheap but indiscriminate and had harmful effects on adjacent fruit crops. New chemicals were therefore developed like Atrazine, which its makers Geigy Chemical Corporation claimed yielded a fourfold return on investment, since one giant foxtail plant per foot of corn was shown to reduce yield from 93.5 bushels/acre to 86.5, and Atrazine yielded 19 bushels/acre over an experimental control area.

Fertiliser application rocketed. In 1945, US farms used 419 million tons of nitrogen and 435 million tons of potash. By 1970, they were using 7,459 million tons of nitrogen and 4,035 million tons of potash. Yields increased by ten bushels/acre on fertilised fields. Experimental practice determined whether initial and/or later fertiliser application was most effective. Side dressing with nitrogen was shown to increase yield by 13.5 bushels/acre. Autumn fertiliser was shown to require exact soil temperature. Fertiliser use grew from 43 per cent of Iowa farmers in 1949, to 68 per cent in 1959, and 96 per cent in 1969 and came to represent 39 per cent of total farming costs.

Antibiotics and the growth hormone stilbestrol were added as supplements to cattle feed, claiming a return of 10–19 times its cost in enhanced meat production, until it was banned in 1973. To quote Anderson, 'the modern hog business would collapse without antibiotics'. By 1951, almost all Iowa's farms had electricity, and this allowed the installation of automated cattle feeding and milking equipment. Pigs were confined to sheds. Harvests were gathered with balers and combine harvesters.

The huge agricultural revolution saw yields and total output rise dramatically through the deployment of a complex web of technologies. Food became available and cheap. Large urban populations were sustainable. However, the question then arises as to how far and how often such quantum leaps in productivity can be repeated. In 2011, global agriculture feeds a population of 7 billion people. Can it further increase its productivity and output to feed 9 billion by 2050? The Iowa history also shows the cycle of technology leading not only to productivity increase, but also to resistance and externality, as insects mutated to become resistant to pesticide, and environmental, safety and health implications emerged which then constrained the technology.

In February 2011, the Economist magazine produced a review of world agricultural technology and productivity, leading with the question of whether agriculture would be able to feed a world population of 9 billion expected by 2050. The report made numerous strategic observations, including

- At the UK Rothamsted research station

- a low technology wheat field yielded 1–2 tonnes/acre
  - a 'Green Revolution' field yielded 4–5 tonnes/acre
  - a best practice field yielded 10 tonnes/acre
- The difference is entirely due to the technology mix of plant type, herbicide, fertiliser and husbandry.
- Humans need 2,100 calories a day, and 90g of meat a day, according to the United Nations Food and Agriculture Organisation and the Lancet medical journal. Agricultural production is almost double these requirements on average.
- The meat, dairy and vegetable component of demand will rise from 20% of all calories consumed in 2000 to 29% in 2050, requiring meat production to double to 470m tonnes by 2050
- Growth in crop yields has slowed from 3% in the 1960s to 1% in 2010 but world population growth is 1.2%/year
- Rice and maize production will therefore have to grow by 1.5%/year and wheat production by 2.3%/year
- Maize planting has risen from 40,000 plants/hectare in 1960 to 90,000 plants/hectare in 2010
- Research spending on maize has been $1.5 billion/year, four times that on wheat research
- There is almost no research spending on African sorghum and cassava, despite evidence that new semi dwarf sorghum has a three times greater yield
- The International Food Policy Research Institute estimates that climate change could reduce cereal crop yields by 9–18% by 2050
- Land, water and fertiliser are the constraining factors of production. Of these
  - Land is not a constraint since according to the World Bank, a further 0.5 billion hectares is available in Latin America and Africa, in addition to the 1.5 billion hectares currently under cultivation globally
  - Land use needs to improve
  - Western Europe yields 9 tonnes/hectare of wheat compared to 2–4 tonnes/hectare in eastern Europe
  - Ghana uses only 3% hybrid maize compared to 90% in Brazil
  - Brazil farms only one head of cattle per hectare which it could double
  - Water is a constraint since according to Nestle, global consumption of 4,500 cubic kilometres/year exceeds the 4,200 cubic kilometres/year available, so that water tables are falling. Meanwhile agriculture will need 45% more water by 2030. Technologies like targeted drip feed irrigation, and no or low till agriculture will be needed.
  - Fertiliser usage in Africa is only 10Kg/hectare compared to 180Kg/hectare in India so that more fertiliser would raise yields, but fertiliser prices have increased dramatically
- 30–50% of all food produced is wasted due to poor storage silos, roads and refrigeration in the distribution system. The very fragmented, high employment, low productivity, small stallholder food retail structure in India, and the political opposition to supermarkets, leads to

under-capitalisation and extensive waste of food through lack of storage and refrigeration.
- The report concludes that whilst food can be sufficient for a world population of 9 billion by 2050, this will need the deployment of further agricultural technology, principally genetically modified food. Only maize is currently achieving the 1.5% annual yield increase to meet the requirement, and this is due to genetic modification. Wheat, rice and soyabean will have to do the same.

Meanwhile protestors destroy research crops of genetically modified (GM) foods before scientific research on them is complete, showing that social constructivist theories of technology can triumph over science determined technology if the pressure groups are sufficiently determined, forceful, or even violent.

3.2.3.4 Propulsion technologies

Humanity, economy, and society are defined in space time. Although the great majority of people spend the great majority of their time in a very small space of a few miles radius, many people also like to travel considerable distances to see other people, to visit other places, or to work in large organisations operating across huge geographies. Nature offered only human walking, running, and horse transport, whose slow speed and need for regular rest and feeding, restricted humanity, its economy and society, to a small locality. A major shift in the pattern of human life, and in the space time location of its economies and societies was enabled or driven by propulsion technologies. Even when human beings did not move location, the global economy required constant mass movement of products from producer to consumer locations.

The main technologies which emerged to facilitate higher speed movement, increasing the distance and decreasing the time of human space and time existence, were the steam engine, the internal combustion engine, the electric motor, and the gas turbine. But these alone were only the means of power to drive a vehicle. Other technologies were needed to enable the production of a moving vehicle, and yet other technologies to effect the infrastructure for the vehicle to move along. The convergence of very diverse technologies within a technology complex necessary at this macro level of engine, vehicle and infrastructure is already evident. Further than this, the convergence of micro-detailed technologies within each module of engine, vehicle and infrastructure is a highly complex web.

As early as 1690, inventor Denis Papin developed a rudimentary steam engine. This was followed in 1698 by Thomas Savery's steam engine water pump and in 1712 by Thomas Newcomen's more effective steam engine water pump. In the late eighteenth century, James Watt improved on this design with a separate condenser and an engine using 75 per cent less fuel than Newcomen's engine, and in 1800 Richard Trevithick developed a high pressure steam engine. The water boilers for these engines were usually coal-fuelled, and the steam engine became the main traction for the world's railways until the mid-twentieth century. The factory at Votkinsky in Russia was the largest manufacturing plant in Europe in the nineteenth century, and produced a total of 630 steam engines. The composer Tchaikovsky's father was the General Director of the plant from 1837 to 1848, and it was here that Tchaikovksy was born in 1840. It was a century later, in 1938, that Sir Nigel Gresley's 'Mallard' steam engine – built in Doncaster, UK – achieved the world speed record for steam locomotives of 125.88 mph (202.58 Km/hr). Meanwhile steam-powered ships carried goods and passengers around the world. The SS Great Britain, launched in Bristol, UK in 1843, pioneered a new generation of technology, with a 1000 horse-power (HP) steam engine driving screw propellers in an iron hull. Diesel engines and hydrojets later displaced steam as the propulsion power for ships.

The internal combustion engine powers the ubiquitous motor car. After the US Electric Vehicle Company went bankrupt in 1907, Henry Ford's mass produced model T, 'you can have any colour as long as it's black', initiated mass production of the car. According to one study, there were over 600 innovations in the US car industry between 1900 and 1981, including hydraulic brakes, front-wheel drive, air conditioning, power steering, fuel injection, and electronic ignition. The Volkswagen Beetle achieved the highest total production of any car with over 20m eventually produced worldwide, until this was overtaken by the Toyota Corolla with 35 million cars produced by 2007. The car provided a driving force and seed bed for a wide range of technology development. Initially the objectives of automotive technology were to increase performance measured in terms of speed and acceleration, enhance the functionality of creature comfort, increase efficiency, and lower cost. A later ecological focus was to reduce emissions and fuel consumption. Both favoured lower weight and higher power/weight ratios which led to efficient wind drag designs, and hence a rather standard shape, and the use of light high strength/weight materials. Aluminium replaced steel in car engines, initially for the engine head, and later for the whole engine block. This drove a need for piston cylinder wall strengthening technologies with hypereutectic 17 per cent silicon aluminium alloys, or high resistance plasma sprays. It also led to advances in aluminium casting technology, achieving near net shape castings to avoid porosity and reduce machining cost.

Magnesium competed as a strong light material with the additional benefit of its sound reduction function. Volkswagen Beetle engines had originally been made of magnesium, sourced from the deposits at Porsgrunn in Norway, made more readily available by the Nazi invasion. Technology therefore had a major political and social effect. Magnesium had been extensively used in helicopter gearboxes, but its corrosion in an electrolytic cell with steel and water led to it being abandoned in this application until Norsk Hydro developed new AZ91D alloys capable of withstanding this corrosion. It looked to other applications markets, from the loudspeaker chassis for its sound deadening qualities, through sports crossbows to a laptop computer chassis, where its strength/weight ratio gave an advantage. Magnesium also had an advantage over aluminium in high pressure die casting applications in that it adheres to the steel die less, and so increases die life considerably compared to aluminium. The economic test for magnesium technology was therefore whether its higher primary cost was more than compensated for by these advantages. Here, the market was determining the technology.

Copper was displaced by aluminium for car radiators until new copper/cerium alloys allowed very thin wall copper radiator specifications which could compete against aluminium again. Otherwise copper fed into electric wiring, plumbing and roofing applications markets. Other technologies developed for the automotive application included: toughened glass for windscreens, catalytic converters for exhaust emission cleaning, advanced lubrication oils etc. In 1971 Honda launched its CVCC, 'Compound Vortex Controlled Combustion' engine which was the first engine to comply with the US 1975 environmental emission requirements.[107] This led to a significant market share for Honda's Civic car, demonstrating how a combination of ecology and market could determine technology.

In these cases of competing automotive and materials technologies, the market criterion which determines which technology will predominate is the criterion of **price/performance**. This may not be a totally determining objective criterion, since there may be information distortions and inadequacies. The example is often quoted of how VHS technology won against the competing β-max technology for television video players and recorders. In this case, β-max was the superior technology when measured by the market criterion of price/performance, but extremely determined market promotion succeeded in pushing VHS technology to the fore. VHS quickly became the standard in an application market where only one standard could operate, since interoperability of video cassettes with video players was essential. As in all markets, information distortions, sometimes created by overwhelming advertising promotion, can determine outcomes.

Another challenge to the determination of technology by its competitive price/performance rating is the example of word processor and spreadsheet software. Currently, Microsoft totally dominates the global market with its Word and Excel products, but this outcome is not at all the result of best price/performance ratings. Indeed, historically competitive products such as AmiPro, Wordpro and Locoscript – the latter of which was ready installed on the original Amstrad word processors – were superior products to Word. Locoscript usefully offered multiple clipboards, and AmiPro was visually more appealing. Nevertheless, huge feudal power exercised by Microsoft eliminated these competitors with an inferior product. Word's bugs persist – for example after using a find/replace function on a file, the pagination scroll option using Control+page commands is lost. The same is true of Microsoft's e-mail applications. Outlook Express fails to open historic e-mails once the stored list has become large, and Outlook hangs if the user tries to open an attached file immediately from a fresh incoming e-mail. Microsoft does nothing about this, but continues to update its software to new versions, incorporating the old faults but changing some detail which forces the whole world to update in order to communicate to a common standard. The accumulation and abuse of market power can thus challenge the rule of price/performance, which is why competition authorities should make more effort in regulating against the development of monopoly power. They have belatedly started to do this in the browser market, forcing Microsoft to openly offer users alternative competitive web browsers.

Random factors also enter into technology market outcomes as they do in all contexts. However, as purchasing and procurement skills evolve in a global market characterised by ever more accessible information, the claim that price/performance will determine technology outcomes in a market economy is credible. This insight develops the case we will argue in the concluding part of this book: that technology is not a single determining artefact, but operates in interactive networking with other artefacts, principally that of the market. Technology is market contingent.

Meanwhile the electric motor displaced steam and diesel as the predominant propulsion technology for railway traction. It was only in 1955 that Japanese producers developed AC motors for train propulsion but by 1964, the famous Shinkansen 'bullet train' commenced operation between Tokyo and Osaka at speeds of 210 Km/hour, rising to 220 Km/hour in 1986 and eventually to 300 Km/hour in new generation trains in 1997.[108] The French TGV commenced operation between Paris and Lyon in 1981 and its successor the TGV Atlantique achieved the world speed record of 515.3 Km/hour in 1990. The later German ICE was a heavier train with a lower top speed but included pressurised cabins to eliminate noise when entering and leaving tunnels. Here geography, essentially therefore nature, determined technology. France's topography is characterised by long flat distances between major cities making high speed beneficial in reducing journey times, whereas Germany has more cities located closer to each other, reducing the advantage of a very high running speed.

The market determination of technology became evident in consumer modal choices. Trains became faster than cars, but are not as convenient. Trains however are slower than aircraft, but more convenient in their city centre destinations. European train operators showed that the consumer market would favour the train if journey times were no longer than twice the aircraft flight time and no longer than half the car journey time. This market trade-off parameter then determined the extent of investment into high speed rail. Again a range of technologies from gate turn off (GTO) thyristors to control the AC drive, through regenerative braking, automatic train control safety systems, to new wheel and bogie technologies were developed to enable the high-speed train.

The propulsion technology of the twentieth century was the gas turbine aero engine. Developed by Frank Whittle in the UK and Hans Pabst von Ohain in Germany, the jet engine displaced the propeller engine for long haul aircraft flight. This led to immense technology development, from the carbon fibre steel developed by Rolls Royce UK for their RB211 engine to power the Lockheed Tristar (which was a calculated risk whose expenditure bankrupted the company but succeeded in getting the company into the US aircraft market against its competitors General Electric and Pratt and Witney), to new chemical engineering technologies, single crystal blade technology, lithium alloy technology, and fly-by-wire control technology for the aircraft itself.

3.2.3.5 Fuel cell technology

Fuel cell technology is an interesting case study which allows us to explore examples of the nature of technology. The original discovery of fuel cell technology was made in 1838 by Christian Schoenbein and William Grove. It was not until 1955–58 that Thomas Grubb and Leonard Niedrach, both working for General Electric, developed the usable 'Grubb-Niedrach' fuel cell, and a year later in 1959 that Francis Bacon built a 5KW stationary fuel cell capable of powering a welding machine, and Harry Ihrig a 15KW stationary fuel cell.

Fuel cells produce electricity, heat and water from an input of hydrogen fuel into a cell constructed of an anode, an electrolyte, and a cathode. They are different to batteries which are closed thermodynamic systems, whereas fuel cells are open thermodynamic systems with fuel flow in and output flows. The essential technology is that the hydrogen fuel is split into protons and electrons by the platinum catalysed anode. The electrolyte only allows the protons to flow through it, so the electrons convert to an electric current at the nickel anode. When the protons reach the cathode, they react with oxygen to produce water and heat.

Research and development of fuel cell technology has yielded five types of fuel cell, distinguished by differences in the electrolyte technology. These are listed in the table on the facing page.

As a source of electric power, fuel cells have potential application both in power generation – where they produce direct current (DC), which needs to be put through an inverter to produce AC power for regular transmission, distribution grids and user interfaces – and in motor vehicles, where they drive an electric motor. They therefore have to compete with fossil fuel steam- and gas-turbines, nuclear and wind in power generation, and with the internal combustion engine in the motor industry. The advantage of fuel cells is that they have almost no emissions, are totally quiet in operation, and that they produce no hazardous waste. Since they contain no moving parts, their reliability in operation is very high. Their problem is cost. The installed cost in 2002 was $1000/KW. By

Table 3.2 Fuel cell technologies

Fuel cell type		Efficiency	Operating temperature	Characteristics	Applications
AFC	Alkaline fuel cells	60%	80°C	Established technology Needs pure fuel	Space programme Large scale power generation
PAFC	Phosphoric acid fuel cells	55%	200°C	Large, heavy Slow warm up >90% efficiency for CHP	Large scale power generation
PEM	Proton exchange membrane cells Polymer electrolyte fuel cells	55%	80°C	Low temperature Low weight Fast start up	Automotive Portable power Distributed power
SOFC	Solid oxide fuel cells	50%	1000°C	Very high temperature Can use other fuels	Small and large power generation
MCFC	Molten carbonate fuel cells	55%	650°C	Very high temperature Corrosive electrolyte	Large scale power generation

2008 UTC Power in the US which claims several independent power provider installations worldwide, offered its 400KW power generation unit for $1m installed cost. Companies like Ballard Power Systems are seeking to reduce installed cost by using lower cost carbon silk catalyst and Solupor membranes. For automotive use, comparable efficiency is measured both as tank-to-wheel and includes the fuller value-chain effect of powerplant-to-wheel. Whilst a standard diesel car driven to a standard cycle achieves 22 per cent tank-to-wheel efficiency, Honda's 2008 FCX Clarity car achieved 60 per cent tank-to-wheel efficiency. However in May 2009, the US government discontinued funding for its Hydrogen Fuel Initiative Programme, since other technologies are thought more promising and the issue of a distribution infrastructure for hydrogen fuel looked challenging. Nevertheless, many prototype fuel-cell-powered buses, cars and motorcycles have been produced and tested from the 2001 Chrysler Natrium bus to the 2005 British firm Intelligent Energy's motorcycle. Prototype fuel-cell-powered light aircraft, submarines and boats have also been developed. UTC Power is working with BMW, Hyundai and Nissan, Intelligent Energy is working with Peugeot. Rolls Royce is investing in SOFC fuel cell development, as is Sulzer Hexis. Ballard Power Systems fuel cells are powering buses in Norway. In January 2011, Mercedes Benz launched three of its B class fuel-cell powered cars on a 125 day, 30,000 Km, 14 nation driving trial.

Fuel cell development is therefore subject to

Table 3.3 Fuel cell technology dependencies

Technology subject to	Via
the objective nature of the technology	the feasibility and availability of what can be engineered
human intentionality	the search for cost reduction through lower cost components
the market	the competitive delivered cost of power compared to steam turbine and internal combustion engine commercial decisions to invest in fuel cell R&D
nature	the need to reduce emissions to atmosphere
society	the value of near zero emissions and near zero operational noise to be applied to the market's economic equation
government	the decision to fund the technology in development and application

We therefore see again the combination of factors which determine technology development and application. Technology operates in a systems network model, not unilaterally or bilaterally with any one other entity.

3.2.3.6 Medical technologies

Medical technologies have had a profound impact on human life. Antibiotics, which became available after 1945, attacked bacteria and cured infections, many of which would otherwise have been fatal. Other pharmaceuticals were developed to control human body organs such as the heart, or to counter body chemicals such as uric acid, which was found to be responsible for gout. Analgesics reduce pain, antipyretics reduce fever, anaesthetics became central to facilitating surgery. The use of antidepressants and NSAIDs (non steroid anti inflammatory drugs) became widespread. Antivirals were developed to combat HIV and other viral infections. As bacteria became resistant, so antibiotics evolved to maintain effectiveness. Diseases like smallpox have been eradicated, but malaria is proving more difficult to overcome – the World Health Organisation 2010 Malaria Report estimates that of 225 million cases of malaria in 2009, 781,000 were fatal, mostly in young children. The report states that whilst $1.8 billion was spent on worldwide malaria control through technology strategies such as insecticide treated mosquito nets, an annual expenditure of $6 billion is needed to eliminate malaria by 2015. Here is a clear case where technology alone cannot achieve a result; it has to be combined with the human intent of financial investment to achieve the necessary deployment.

Apart from pharmaceuticals, medical technology includes physical equipment and specialist skills, for example in surgery which has evolved into micro so-called 'keyhole' surgery. Eucomed, an organisation which represents the European medical technology industry, reports that 11,000 companies in Europe work with medical technology, employing 435,000 staff, and producing 500,000 products, with average product life being a mere 18 months. These are classified into 16 product groups as shown in the table overleaf.

The European market for these technologies has a value in excess of €60 billion. Structurally, the medical technology market is differentiated between pharmaceuticals and these other technologies. The global pharmaceuticals sector is very concentrated and characterised by a very small number of multinational corporations due to the huge R&D resource required, whilst the sector for other medical technologies is fragmented and characterised by SME suppliers. This again demonstrates the complex web of technologies that feed into any single application, as well as showing the complex industrial development, production, and market organisation of resources necessary to deploy such technologies.

Table 3.4 A typology of medical technologies

#	Technology	Examples
01	Active implantable technology	Cardiac pacemakers, neurostimulator
02	Anaesthetic respiratory technology	Oxygen mask, anaesthesia systems, breathing circuit
03	Dental technology	Dental amalgam, cement, and instruments
04	Electromechanical technology	Dialysis systems, electrocardiographs, endoscopes
05	Hospital hardware	Hospital bed, baths, sterilisers
06	In-vitro diagnostic technology	Blood glucose monitors, test kits
07	Non-active implantable technology	Cardiovascular clips, heart valves, bone prostheses
08	Ophthalmic and optical technology	Contact lenses, ophthalmic test instruments
09	Reusable instruments	Various surgical instruments
10	Single use technology	Syringes, needles, gloves, catheters
11	Assistive products for persons with disability	Wheelchair, crutches, hearing aid, artificial limbs
12	Diagnostic and therapeutic radiation technology	MRI systems, X-ray
13	Complementary therapy devices	Bio-energy mapping systems, magnets
14	Biologically derived devices	Tissue heart valves, natural grafts
15	Healthcare facility products and adaptations	Safety systems, generators, sanitation products
16	Laboratory equipment	Microscopes, centrifuges, pipettes

The result of the deployment of medical technologies on human life is clear from the graphs on the facing page.[109]

The data on infant mortality and life expectancy is itself startling. Added to this, the choice to reproduce or not through contraception and fertility technology affects the human population quantitatively, and data cannot easily measure and report the effect on quality of life of medical technologies, for example in pain relief. It is clear that technology hugely affects humanity.

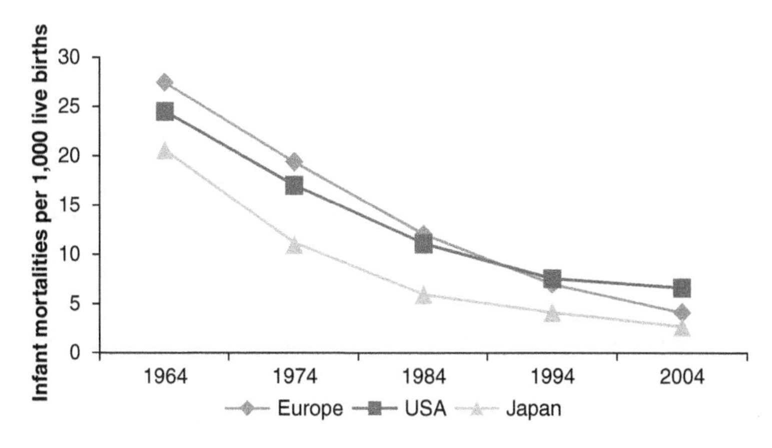

Figure 3.3 Declining infant mortality, 1964–2004

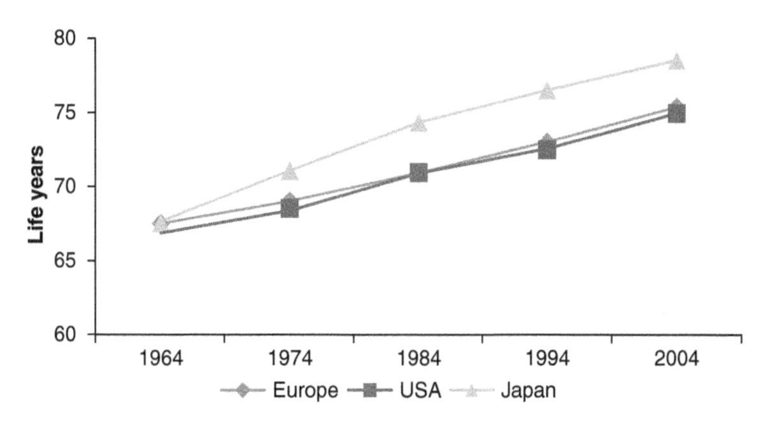

Figure 3.4 Male life expectancy at birth, 1964–2004

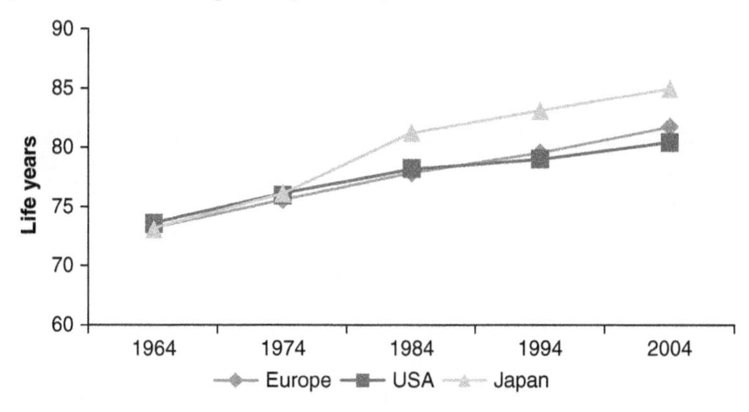

Figure 3.5 Female life expectancy at birth, 1964–2004

3.2.3.7 Systems technology

Whilst some technologies can be seen as an extension of human physical power, systems technologies are an extension, or a multiple automated implementation, of human intellectual capability. Homeostatic control systems with a full feedback loop require sensors to measure condition parameters, processors to monitor condition against requirements, and activators either to correct system settings automatically, or to prompt a human operator to set the correction. Simple examples are thermostatic control of heating systems where the thermostat measures achieved temperature, compares this to required temperature, and then activates valves to switch boilers or other heat sources. Huge pipelines covering vast remote geographies are controlled by similar telemetry systems.

In standard commercial applications, the capability of integrated computer systems to

- take customers orders on-line
- check these against a stock control module for product availability
- organise physical despatch and delivery
- monitor, optimise and manage the depot and vehicle fleet operation to the detail of exact picking lists and vehicle schedules
- generate a customer invoice
- feed through to company cost and sales accounts, and
- update the stock control model to generate new supplier orders when stock levels reach re-order points

is impressive. It implements a level of productivity which releases human labour to higher standards of living, if wider creative activity or new job creation programmes are developed.

Such systems comprise computers capable of holding and processing large volumes of data, together with communications technology able to transmit data over any distance. These systems therefore require a convergence of several technologies: robust and well-calibrated sensor technology, computing technology, communications technology, and physical activator technology. Technology itself is unlikely to be able to orchestrate its own convergence: it requires the intervention of human intentionality, and usually of the market artefact. Since complex convergence of individual technologies is pervasive, widespread, and almost the norm, it argues strongly against a view of the autonomy of technology, since as we have shown such convergence of technology requires human or market agency.

There are debates about how far computerisation has raised economic productivity. The standard contemporary experience in a retail shop sometimes suggests that computerised sales-order processing and stock control requires far more time from the operator than a simple cash purchase with no information recording. But that observation obscures the background benefits to the wider supply chain operated by the retail company. Information capture can often be an immediate efficiency loss, but the wider systems benefit is the more important efficiency criterion. Nevertheless, the intrusion of integrated information technology systems into many if not all areas of life does alter the human experience, making interactions more technical, more remote, essentially less human. Or the experience is changing what we understand the definition of 'human' to be.

In large industrial organisations, whole armies of clerical workers were displaced by computer systems. IBM introduced its 360 mainframe computers in 1964. Its main developer, Gene Amdahl, later left to establish his own competitive 'plug compatible' mainframe computer company. These mainframe systems were widely introduced to operate commercial companies' management systems, usually starting with accounting modules, then extending to personnel, then to include sales and purchasing operations. These computers were often custom programmed for each user, using the COBOL commercial programming language, again showing the importance of technology convergence for technology progress. Meanwhile the Digital Equipment Corporation's famous PDP and later VAX minicomputers were widely applied to scientific and manufacturing production environments, usually programmed in the more mathematical FORTRAN language. Such systems enhanced productivity both by reducing the human labour cost of many manufacturing and commercial functions, but also by introducing higher and more consistent standards of accuracy which proved crucially important in raising product and service quality.

Initial implementations were fraught with difficulty. Equipment systems technologies, and therefore important data, were insecure: operators were required to mount magnetic tapes, for example of a company's product, material and work in progress stock, and the careless operator who instead mounted a completely blank tape caused the system to issue immense supplier orders. Scientific and engineering calculations took many hours to accomplish on this original mainframe technology, and users generally had to hand their calculation 'run' to specialists in white coats operating the computer who would deliver a single output a day later. User interaction was not allowed.

Data was input on the iconic IBM 129 punch card which was a thin card containing 80 columns each with three header positions, allowing all characters to be expressed. Square holes were punched into the card to record data, and card readers were developed to rapidly read these cards into the computer. Computer program code was also input using these cards, and some programmers developed the exotic skill of being able to read these cards quickly from the square holes punched in them. For more ordinary souls, IBM provided punch card interpreter machines which would print the data held on the card along its top edge. A bug in a computer program required tasking the computer to spill out great wads of these cards so that the programmer could determine the coding error, replace the wrong instruction and recompile the program by feeding the cards back into the computer. COBOL and other programming languages all had their unique compiler program, translating the high level language instructions into binary machine code. Those, as they say, were the days!

It has, however, been claimed that the period of rapid productivity gain due to widespread computerisation of everything (1965 to 1980) was followed by a 'productivity paradox' where further computerisation benefits were diverted into ever greater overhead management requirements. In this case, the technology may well be determining human practice instead of human requirement determining the technology: they are in genuine symbiosis.

The convergence of computing and communications technology had long been promised. Available technologies for the transmission of data and image, as well as the voice, were hindered in their implementation by the inertia, and often indifference, of large state-sector telecommunications operators. In the UK, simple 1970s technologies for telephone-answering equipment were banned for years, due to the failure of the telecommunications authority to issue the necessary technical accreditation, which was simply due to organisational lethargy. So again we see that technology is not universally autonomous, but contingent. Similarly it was often claimed that large monopolistic computer companies developed technology but then kept it 'on the shelf', withholding it from the market until a time when the company's business cycle and strategy required its release. This is another example of the contingent nature of technology.

Communications technologies made significant advances. Multiplexing and packet switching was fundamental to more widespread telephony by allowing mixed traffic to be separated into packets fed down the same physical cable and decoded for each user destination. Optical fibre technology hugely increased the transmission capacity of cables. Nevertheless, transmission of text was by inefficient facsimile machine technology for very much longer than the newly available technology required. Facsimile was widely installed in the mid-1980s when simpler more immediate coded text transmission technology was readily available. As with voice recording technology, this is an example of where an institutional interaction of government regulation, technology availability, and the consumer/producer market dynamic took an inefficient turn. Society opted for facsimile technology even though a basic version of more effective internet technology was soon to become available. In the network of artefacts in our systems network model of technology, the artefact of society can also therefore exert influence and determine outcomes. The eventual outcome is a result of the combined interaction of all these artefacts of science, technology, market and society, as well as of nature's resource and human intentional action.

When it came, global convergence of computing and communication technology, and the almost global deployment of its result to individuals, was a revolution of the configuration of humanity and technology which further impacted society, the market economy, and the space time location of humanity. E-mail revolutionised personal and corporate communication. More widespread and more frequent communication became possible. Quality played off against quantity to the extent that the infamous SPAM e-mail from advertisers and other unsolicited sources required further technology to block and divert it. People now had blacklists and white lists. Business organisations attached files whether in text form, or as presentations or spreadsheets. Everyone could be copied in and everyone was, leading to a huge editing task most days for professional workers. The technology which had automated the human secretarial function now reimposed it, an example of how technology really can determine social structures.

The ubiquitous web site followed. Organisations became no longer recognised by their physical presence or location, but by their virtual location on 'the web'. But this presence, in which many organisations had invested heavily, was not easily discovered or found. So search engine technology was added, and a new verb emerged into regular language: people everywhere 'Googled' everything. In terms of content, Wikipedia's on-line encyclopaedia was widely consulted. Human awareness became immensely extended by the use of these technologies. Human possibilities and options too. However, the wider spread of the human experience arithmetically led partially to its enrichment and partially to its dilution. Life partly migrated from the real to the virtual domain, and the outstanding question is whether the totality of life is thereby greater or less. By becoming much more virtual in its emphasis, life has challenged both the physicalism of contemporary philosophy, and the incarnation of Christian faith. As with all metaphysics, the virtual is hosted by the physical provided by Cisco servers and fibre optic, satellite and copper twisted pair networks. It cannot exist independently of this physical entity, the so-called philosophical concept of 'supervenience'.

Creative innovations abounded. Hardly had the worldwide web 'community' been established, than the concept evolved further. Social networking was now the lead idea. Many millions of users registered with the Facebook site, uploading and tagging their photographs for the world to see, and extending the concept of friend very widely, and one might argue, very thinly. Now every previously ordinary person could have celebrity status. They became their own paparazzi, publishing previously private details of their lives. Before long the commercial advertising potential of this quintessentially proliferating technology became apparent, and advertisers placed auction bids to get their message to very specifically targeted audiences, in all rendering Facebook worth over $100 billion at its 2012 stock market flotation. Twitter enabled mobile phone users to establish a following, measured for some in millions for celebrities like Stephen Fry, and dignitaries like Queen Rania of Jordan.

3.2.3.8 Ecological technologies

The first wave of the industrial deployment of technology took little account of its environmental impact, or its human impact more generally. Emissions to land, air and water were immense. Landfill was the low cost methodology favoured for waste disposal. The irreversible polymerisation of petrochemicals led to plastic waste streams new to nature, which nature therefore had no process to absorb. Clean air became a scarce rare commodity, London became infamous for its 'pea soup' fog, Los Angeles for its traffic generated smog. Acid rain poured down as sulphur dioxide in the atmosphere dissolved into the raindrops. Populations close to chemicals factories found blotches of black stain on any clothing hung out to dry. Land was extensively contaminated with heavy metals. Spent nuclear fuel had to be stored underground. Health was adversely affected as asthmatic allergies spread. Oestrogens from washing powders entered the water table, leading some to suspect them responsible for a reduction in the human male sperm count. In a helicopter above the Siberian city of Leninskuznetsk it was impossible to see the town or countryside beneath the black cloud of pollution from the emissions of its coal mines which hung over it.

Electricity generation was responsible for much of the sulphur dioxide and sulphur trioxide SOx in the atmosphere, as well as the nitrogen dioxide and nitrogen trioxide NOx pollution resulting from its combustion of sulphur and nitrogen containing coal and gas. Low sulphur coals, particularly Russian coal which came to dominate the European coal import market, helped reduce SOx emissions, but technology for flue gas desulphurisation (FGD) was widely implemented, adding to the production cost of each unit of electricity. The same is true for other low emission power generation technologies. Nuclear power is more expensive than fossil fuelled power, and renewable power arguably the most expensive of all. Lobby groups who effectively opposed nuclear power forced extended fossil fuel combustion which added to emissions: this switch was certainly the case in Germany in the early 2000s. A cleaner atmosphere would come at a price, and so technology, the market economy, and society through its government's standards legislation, had to co-determine the preferred outcome. These three artefacts in the systems network model of technology are again interacting and co-determining outcomes.

Similarly with NOx emissions, governments took various steps to impose limits. In the UK these were in terms of maximum permitted NOx emissions of $500mg/m^3$ from 2010 and $200mg/m^3$ from 2016. The effect of the first step was to require the installation of low NOx burners, and to exclude certain coals, notably South African coal whose combined nitrogen content and volatility yielded NOx emissions in excess of this, from combustion. During this period, the UK only had its own domestic coal, together with Colombian and Russian coal, available for compliant combustion in its power stations. The effect of the more severe limit to be introduced in 2016 will be to require all power stations combusting all coals to implement selective catalytic reduction (SCR) technology, as is commonly already deployed in continental Europe. Electricity generation costs and prices will rise. Meanwhile the Japanese standards authorities limited the nitrogen content of imported coal to 1.7 per cent, restricting imports to Australian, Indonesian and some Chinese coals and excluding Russia's Kuzbass coals with its higher 2.2 per cent nitrogen content. Here is a complex example of technology, cost of production, and environmental emissions interacting and resolving through a combination of further remedial technology, imposed legislative standards, and what became known as the BATNEEC principle – the adoption of the best available technology not entailing excessive cost, a concept which was initially introduced in the 1984 European Union Air Framework Directive.

Government directives on emission limits have to combine several considerations: the technological feasibility of reducing the emissions of an industrial process, the economic investment and operating cost of implementing any mitigation technologies available, the outcome in global competitive markets if competing nations do not insist on such costly emissions reduction for their industrial plants serving the same homogenous product markets, and the degree of urgent need attached to the production and consumption of the product concerned.

This brings us to the central ecological concern of global warming occurring due to CO_2 emissions. In democratic societies, concerns and protests regularly accompany the development and deployment of technology. In this sense at least, technology is socially constructed and politically contingent. The Luddites opposed mechanisation, Gandhi triumphed with basic spinning technologies, nuclear power protest lobbies have been effective, agricultural pesticide, fertiliser and genetically modified crops have attracted forceful protest. All these protests have become somewhat eclipsed by the shared concern that levels of CO_2 in the atmosphere will lead to catastrophic climatic results with sea levels rising to flood low lying countries like Bangladesh. Campaigners insist that the science is established truth. Although this leaves insufficient room for necessary doubt, it is clear that even if the climate change hypothesis has any reasonable likelihood of happening, avoidance strategies are urgently needed. These range from drastic reductions in consumption, travel, and the 'carbon footprint' of everything, to hopes that technology can be relied on to come up with answers which will allow continued high consumer lifestyles with minimal ecological consequence. This latter position mirrors that of those whose blind faith in the market feel able to rely on market mechanisms to solve all humanity's problems and dilemmas. Artefacts create faith in themselves on the part of a credulous human following.

The solutions suggested by technology so far range from quantum step increases in the efficiency of technologies whose implementation contributes most to emissions, to attempts to store the offending CO_2 underground in aquifers or depleted oil reservoirs. The latter offers the best of both worlds in that the increased pressure stored CO_2 would create in these oil reservoirs would in fact generate commercial oil whose market value would underpin the so-called Enhanced Oil Recovery business case for carbon sequestration and storage. It seems rather counterintuitive that pollution can be resolved by being funded in a way which produces oil, which is the commodity initially responsible for much of the pollution the strategy is designed to counter. Meanwhile the market artefact is not quiet on the subject and is having a go at offering its type of resolution by fixing caps on emissions for each polluter and allowing polluters to trade carbon credits; although, as reported in the UK Daily Telegraph in its 30 January 2011 edition, from 2009 to early 2011 this market was the victim of smart scams, estimated in the press to have cost the European consumer some €5billion. The efficiency-from-enhanced-technology route to mitigation of CO_2 emissions includes the materials technologies for ever lighter cars, and the high temperature special alloy boilers with enhanced thermal efficiency for power generation discussed above.

Others will see the environmental issue as a major market opportunity itself. The German engineering firm Siemens claims to be 'the world market leader in green technologies. Its web site offers a range of such technologies including

- wind power turbines
- high thermal efficiency powerplants
- low energy lighting
- high voltage direct current power transmission to reduce transmission losses
- its Meros technology for the reduction of emissions in steel sintering
- electrostatic precipitators for dust removal from waste gas streams
- membrane technology for water purification
- its biochemical Cannibal technology for reducing sewage sludge

The company plans to achieve €25 billion sales in 2011 from this technology portfolio. Its French competitor Alstom offers a similar range of green technologies and has deployed its environmental retrofit solutions to many existing powerplants. The artefacts of market, technology and society are again interacting with human intentionality.

3.2.4 Productivity

The key variable mediating technology to humanity is productivity. Surprisingly, this fundamental truth seems to be little understood, and is paid little attention whether in academia, politics, economic management, or media comment and analysis. It receives as scant regard as technology does in philosophy. Yet technology and its progeny productivity are the joint mechanism which is so noticeably redefining the human experience; and, we can argue, also redefining human ontology, what it is to be a human being, as well as the nature of human society. **Productivity is certainly the key variable in economic theory and economic analysis**, a core role which is rarely understood or acknowledged.

Productivity is quantitative, measured as the number of hours of human work required to produce a standard unit of output. It is not however always easy to measure, because at the same time it constantly redefines the unit of output itself. Hence, whilst the number of person hours required to construct a car has radically declined, the nature of the car itself has grown to offer higher functionality, so that the underlying productivity has in fact grown by a multiple of these two factors. This twin phenomenon is true of nearly every economic product.

Productivity is a real variable, measured in other real variables of output and time. It is not a financial measure. Real productivity causes financial variables and financial measures are the outcome of the real economy, rather than the other way round. Financial prices, the rate of price inflation, levels of consumer and public sector debt, pensions payable, and very definitely standards of living, are all dependent on productivity. If productivity were infinite, prices would be zero; if productivity were zero, prices would be infinite. High productivity growth will lead to low or even negative inflation; low productivity growth will drive inflation up. It is true that financial variables can affect each other; for example an increase in the supply of money given constant real productivity will feed inflation pro rata, but this is a trivial endogenous effect. **The ultimate driver of the economy is productivity.**

Productivity drives both prosperity for society, and profitability in the business model. Economic theory derived from David Ricardo postulates three fundamental resources: land, labour and time. But these three prime resources are then reconfigured to yield the secondary resource of capital, not defined as money, but as real infrastructure, systems, and equipment. Capital is a combination of land and labour accumulated over time. Capital is then engaged in production, so that economic output results from the 'production function', combining capital and labour over time. And **the rate at which capital and labour combine over time to produce units of output is the definition of productivity.**

Technology is also part of real capital in this production function, since 'human capital' is the know-how of a technically educated workforce. So in an equation where x units of physical capital K, combine with y units of labour L, over z units of time T, to produce q units of output O, ie

$$xK + yL + zT = qO$$

then technology is simultaneously

i) incorporated within each unit of capital K, and each unit of labour L
ii) reducing the coefficients x, y and z which determine how many units of capital, labour and time are required to produce one unit of output O
iii) increasing the functionality of each unit of output O

Productivity has grown exponentially in very short recent history. Vaclav Smil rightly calls this a 'saltation', – a quantum evolutionary leap compared to Darwinian micro level mutations. Scattered through Smil's text[110] are various eclectic but representative examples, eg

Between 1900 and 2000

- global consumption of fossil fuels and primary electricity expanded 16-fold
- the global population nearly quadrupled from 1.6 to 6.1bn
- global average per capita supply of primary energy more than quadrupled

- space heating units delivered five times more heat output per unit of input
- US agricultural employment fell from 35 per cent to 5 per cent of total employment
- US service sector employment rose from 31 per cent to 80 per cent of total employment
- a lumen of electric light became 4,700 times more affordable

Between 1940 and 2000

- US employment in manufacturing, mining and construction fell from 47 per cent to 19 per cent of total employment
- US industrial output rose 11-fold
- US per capita GDP increased by a factor of 8, Western European per capita GDP by a factor of 6, and Japanese per capita GPD by a factor of 17
- computers calculation power rose from one flop (floating point operation per second) to 35.86 teraflops.

By 2000

- international phone calls were made automatically, whilst in 1915 the first such call required five operators working for 23 minutes
- in the US, an hour of factory labour generated 4.2 times as much value as in 1950
- the world had
    - 2 billion radios
    - >1 billion telephones
    - >1 billion televisions
    - 700 million cars
    - 30 million Km of roads
    - 1million Km of electricity lines
    - >10,000 aircraft flying at any one time

The twentieth century, Smil concludes, 'experienced economic growth that was unmatched in history'. He shows how quality of life, measured by parameters such as infant mortality and life expectancy, is dependent on per capita energy consumption, with a minimum annual need of 1.2 tonnes oil equivalent, and an optimum of 2.6 toe; above which, Smil claims, there is no further increase in the quality of life, despite which the US per capita consumption in 2000 was 8 toe. Counterbalancing the beneficent effect of productivity, Smil also points out that in 2000, 1.26 million people died in road accidents, 90 per cent of them in low- and middle-income countries.

Smil's account conflates the undeniable increase in productive output with the increase in productivity, so that the question arises to what extent productivity was the key factor responsible for output growth. The American economist Robert Solow in his 1957 paper 'Technical Change and the Aggregate Production Function' applied econometric analysis to conclude that between 1909 and 1949, output per person hour doubled, and that 87.5 per cent of the increase was due to technology, and 12.5 per cent to the increased use of capital. In his 1985 paper, Edward Denison subsequently refined this to claim that between 1929 and 1982, 55 per cent of US economic growth was due to know-how, 16 per cent to improved resource allocation, and 18 per cent to economies of scale, effectively allocating 89 per cent of economic growth to technology in some form of application. Smil is less persuaded by quantitative econometric analysis and prefers qualitative description. He cites the case of the $1.3 million investment in Fairchild Semiconductors in 1957 which led to the foundation of Intel in 1968 and the subsequent founding of over 200 companies including AMD and National Semiconductor. Such technological proliferation depended in Smil's view on education, accessible venture capital, innovation, risk-taking, an enabling legal framework, and protection of intellectual property rights.

Technology-led productivity radically altered the economy, and the focus of economic analysis. Prior to the exponential economic growth enabled by productivity, human demand was chasing inadequate supply; following the boom in productivity and therefore in output, Keynesian demand management became necessary to ensure adequate demand to consume available supply and maintain full employment.

3.2.5 The economy

3.2.5.1 Forms of economy – command and market economies

An economy is an artefact which organises output production of goods and services from input factors of raw materials, labour and capital. In its advanced form, it does this by the intermediate step of investment in an infrastructure of factories, transport, communications and other network systems, and investment in human beings themselves, to increase their capability through education and training. The output of goods and services is for consumption – as the economist Adam Smith famously said 'the end of all production is consumption'.

An economy can be of various forms. **Command economies** operated under socialism in Russia from 1917 to 1990 and in Eastern Europe from 1945 to 1990, and still in North Korea in 2011. Massive central organisations, like Gosplan in Russia, attempted to decide consumer needs, and organise investment and production through Leontief input-output matrices. The coefficients of these matrices represented the productivities of the economy, the rate at which inputs were translated into outputs. Given the complexity of the modelling task, these coefficients were fixed rather than flexible, so that the application of the model militated against growth in productivity, whether this was from exogenous developments, or endogenous to the economy. This technical point was undoubtedly partially responsible for the failure of the USSR economy to enjoy the productivity growth experienced in western market economies in the same time period. Economic management was quantitative, rather than based on money values as in western market economies.

This also resulted in massive inefficiency, as factory management became quantitative in every way of thinking. The more steel a plant was allocated by Gosplan, the more privileged the factory felt. As a result, products contained far more steel than was necessary. Castings, for example, were always very thick walled, whilst competitive western markets drove for efficiency, in this case with thin wall castings strengthened with alloys such as steel/manganese, vanadium, chromium, hypereutectic aluminium/silicon, copper/cerium etc. Plant managers boasted of the high quantitative level of their production without considering quality. In one case, a factory in Druzhkovka, Ukraine, producing high load hydraulic roof supports for the Soviet coal mining industry, produced the enormous output of 32,000 supports annually, far greater than any western plant. But this was only because their life was less than a year due to deficient hydraulics, compared to a life of up to ten years for the western product. The same applied to a conveyor plant in Karkhov, Ukraine, whose annual output of 2,100 was entirely due to the lack of any case hardening of the conveyors' drive gears, again making the conveyor's life short and its replacement frequent.

The application of techniques such as linear programming, whose Simplex algorithm was devised by the Russian mathematician Leonid

Kantorovich in 1939 from his work in optimising plywood production, and further developed by George Dantzig in the US in 1947, flourished in command economies in their 'top down' attempt to organise production efficiently. Kantorovich indeed won the 1975 Nobel Prize for Economics and was the only USSR economist to do so.

This was a very appropriate award, since the simplex algorithm was a major breakthrough in mathematics. It allowed maximisation of linear functions, whereas previously only exponential functions with independent variables squared or raised to a higher power had been soluble, using differential calculus to identify the maximum value of a function (by setting the first differential equal to zero, and ensuring that the second differential was negative, and the maximum was not just a local one). It spawned the whole academic science and professional application of Operations Research, with its deterministic and heuristic algorithms to optimise a wide range of systems phenomena. By the 1960s Moscow had become the world centre for huge operations research models, and practitioners from private sector companies and public sector agencies worldwide flocked to learn from the Soviet experience in OR, in order to then apply the technique to their industry. The famous 'transportation algorithm' flourished in extensive applications.

In 1964, two inspired researchers at the UK Cooperative Wholesale Society in Manchester, G Clarke and J W Wright, developed their famous heuristic 'savings algorithm' which would schedule delivery vehicles from a depot in a way that covered the least total distance, or deployed the smallest vehicle fleet, the latter being the most significant cost driver. Their algorithm was published in the Operations Research journal, coded by the computer firm IBM, and sold as their VSPX (Vehicle Scheduling Package Extended) software package. IBM's main aim in this was to sell more of their then 'Series 1 minicomputers' on the grounds of the claimed 10–15 per cent saving this could effect in a company's secondary distribution cost. A small UK software company, Analytical Systems Ltd, enhanced this algorithm with its proprietary 'look ahead' algorithm. In 1968 ASL successfully implemented the first application of the technology to the daily scheduling of BP petrol tankers to filling stations for Pakhoed in Rotterdam, using a supplementary full-enumeration algorithm to optimally allocate different fuel types to tanker compartments. The lorry drivers were aghast – they had by then become used to computers calculating and paying their wages in the early adopter technophile Dutch economy, but were astounded that they could also route their lorry tankers! The technology is still being further developed, for example by Doyuran and Catay in their 2011 Journal of the Operational Research Society article 'A robust enhancement to the Clarke-Wright savings algorithm', and such algorithms are now in common use by anyone using tools such as Google Maps to obtain journey directions. The technology was also later integrated into sales order processing and stock control systems to effect integrated logistics management for major company users.

In the exciting and inspiring era of the 'white heat of technology' proclaimed by western leaders such as the UK Prime Minister Harold Wilson in the 1960s, it appeared that OR was a tool to ensure optimal operation of everything. Technology could organise and rule for humanity, in the way in which the competing artefact of market was claimed to do in a later epoch, this time trumpeted by another UK Prime Minister, Margaret Thatcher. Indeed, this competitive dynamic between technology and market, as alternative artefacts offering to organise human life, is a key issue we return to at the conclusion of this book.

At the time though, it seemed that any objective function could be expressed mathematically, all constraints could be programmed, and hey presto, with the press of a button – albeit with a rather long wait for the computers of the day to work through the calculation – maximum cost efficiency and maximum fulfilment of all objectives could be assured. Like all humanity's attempts to hand responsibility to a non-cognitive agency, running from the Delphic oracle onwards through other religions, now technology in the form of operations research mathematics, and later the much celebrated market artefact, this was doomed by excessive expectation. These huge models were soon seen to be 'black boxes'. Humanity was expected to simply accept their instructions, but could not engage with their inner calculations, and therefore could not interact with the process. It became clear that most realities were too complex to be fully included in such models, and that efforts to extend the models to wider reality rendered the models too complex to manage or compute. Moreover, the models were unable to incorporate human choice, or the pervasive probability which was found to characterise many systems in a stochastic rather than deterministic world. The models morphed into more interactive decision aids such as 'visual interactive modelling', so that the technology effectively reverted to the role of servant to cognitive humanity, rather than assuming the master status of the black box OR technology. It is thus an interesting case study of technology playing the master/servant game with humanity, an issue which is central to the concluding sections of this book.

However it was political dogma and initial ethical considerations which underlay the operation of command economies, since socialism decreed that distributive justice required state ownership of capital assets, the 'means of production' as defined by Karl Marx. Such a policy was also adopted by the British Labour Party in its (in)famous Clause IV, drafted by Sidney Webb in 1917, the year of the Bolshevik revolution in Russia, and only discarded at the instigation of Tony Blair in 1995. But the communist economic system became sclerotic and atrophied, since it removed and even castigated individual human incentive and initiative. It proved impossible to-control economic activity effectively. Freedom and scope for private endeavour were shown to be important aspects of an effective economy, which needed to operate 'bottom up' rather than 'top down'.

Command economies were essentially production oriented and production dominated. The worker was the heroic figure, there were no shareholders other than the state, and the customer was reckoned lucky to receive the limited goods and almost no services the system offered. Rationing by queue was common. Crowds surged around vehicles making bulk deliveries to dismal shops. Huge quantities of one design of glass vase or anorak or sporting trousers temporarily filled the shelves which were soon bare again. There was no concept of customer service: indeed, hangovers from the command economy, such as the former Russian state airline Aeroflot, even today find customer service an impossible concept and treat their passengers with indifference at airport counters. The economy was not only production oriented – it was military in its focus and priorities. Consumer goods were allocated for production, almost as an afterthought at essentially military factories such as the factory near Tula in Russia which made launch vehicles for intercontinental ballistic missiles, roof supports for underground mining and, as its afterthought, 200,000 children's bicycles and a million steel quadrant bread bins each year. Even in the post-communist 1990s, producers could not adapt to the concept of consumer sovereignty. Paint manufacturers such as Lakokraska in Yaroslavl, who had seen the Finnish company Tikkurila take a huge share of the Moscow market for decorative paint, could not accept that they should deliver to and serve the consumer market, but fully expected eager customers to drive lorries to Yaroslavl to take whatever paint they were offered. Similarly the abrasives manufacturer Urals Abrasives in Chelyabinsk refused to deliver its abrasive grit to an interested Austrian customer, but told the customer to send its lorry to Chelyabinsk to collect the product.

Technology was the responsibility of technical institutes, each focussed on one industry sector, and usually located in Moscow, or in Akademgorodok, a special scientific research town established with its own huge artificial lake 40Km from Novosibirsk in western Siberia and originally instructed by Nikita Kruschev to develop designs and construction technologies which would equal American skyscrapers. For example, the coal-mining technology institute Uglemash in Moscow developed longwall underground mining equipment capable of working the thin coal seams of the Ukraine, which were often 1Km underground, about 0.5m thick and frequently nearly vertical. Such 'heroic' achievement was purely technical rather than economic, as no other world economy would have sought to exploit such fundamentally uneconomic coal seams. Technology in the command economy, like production, was purely quantitative engineering, and had no concept of comparative economic value. The lack of any economic incentive, or even measurement of value by a price signal, together with the lack of social and political freedom to develop technology, meant that it regressed, and the technical institutes were often reduced to reverse engineering copies of western products. The Zhuguli car produced in quantity at Togliatti near Samara was built with an imported second-hand production line designed for Fiat's 128 car, and the Volga saloon, built in smaller numbers at Nizhni Novgorod, was a copy of the UK's Vauxhall 101 car.

By 1990 it was obvious to socialist state television viewers – watching westerners driving Mercedes and Golf cars compared to their Russian Zhigulis, or the amazing (now iconic) East German Trabant, with its cellulose body and massively polluting two-stroke engine, for which grateful customers had to wait 15 years – that the command model had relatively failed and fallen behind the productivity of western market economies. Indeed, the only way these command economies could retain their citizen consumers was to lock them into their countries and deny them freedom of travel or abode. This was sold as protecting the socialist citizen from the insecurities and errors of the west. Western supermarket shelves were said to be full because western consumers could not afford to purchase their goods. Russian doctors decreed western pharmaceuticals and healthcare products, such as sanitary towels and tampons, to be unhealthy. But in fact the virtual imprisonment of the whole population was enforced by the brutal heel of military might, to which command economies diverted their resources. Stalin's terror was history, but it had developed a mindset. The situation paralleled that in the Bible's Exodus story, where Moses had been able to lead the people of Israel out of Egypt, but getting Egypt out of the people of Israel then proved more difficult.

Market economies characterised the rest of the world, although these operated under various degrees of freedom or constraint, of individual initiative or government planning. In most cases, regulation of standards was considered important for a fair and effective economy. Structural regulation also ruled against monopoly and cartels, since the leading concept was for the economy to be competitive, to ensure full potential production at the lowest consumer price, thus generating higher standards of living for the many rather than for the few.

Market economies generally became money economies, although in their early development barter exchange was prevalent, and indeed the Russian economy experienced a barter phase in the 1990s due to a lack of money in circulation. I personally witnessed a Russian manufacturer of roof supports for coal mining being paid physically in wholesale quantities of sugar, which was then paid as wages to workers who resold smaller quantities into the roadside retail market, either for small cash or further barter. Barter is a very inefficient and ineffective means of exchange, since it requires multiple trading posts, and cannot readily equilibrate a homogenous price in an independent unit. Money therefore became the universal transaction medium but also an artefact in its own right; like all artefacts it developed a virtual independent power to constrain rather than enable the real economy it was supposed to service. Some examples of this perfidy are given below.

Nevertheless, different cultures developed different market economies. The US market economy is more reliant on private initiative and is, as a result, more unequal than the European and Nordic market economies which are more communal – a word which is more apt than the frequent pejorative use of the word 'state', with its Kafkaesque overtones. The French market economy was more *dirigiste,* with free markets operating within overall indicative plans. The Korean market economy is more militarised in organisation through its famed chaebols, whilst the Japanese market economy is more socially ordered and more equal as a result.

What market economies have in common, however, is their use of value signals and indicators rather than the quantity signals of a command economy. Market economies operate through the signals of price and profit. Products which cannot sell in sufficient quantity at a price above their cost will not survive, nor will production companies who fail to make profit and therefore become bankrupt. The 'marginalist school' of economics provided the analytic for the simplest models of a competitive market economy. This analysis defined the equation *profit = revenue – cost* and differentiated it with respect to output. Classical calculus shows that setting this differential equal to zero obtains a local maximum (if the second differential is negative), so that profit is maximised at the point at which marginal revenue with respect to output equals the marginal cost of output. This simple finding was taken to determine the behaviour of the firm in a competitive market economy. Firms are said to seek to maximise their profit and so produce to the point where declining marginal revenue meets increasing marginal cost.

Technically, marginal analysis has some similarities with the techniques of a command economy, since the way the simplex algorithm works in a linear programming optimisation is to find the 'shadow prices', which are the coefficients showing the highest marginal effects on the objective function of a particular causal variable. The difference however is that real actors can see profit and are directly impacted by it, and they can equally calculate or experience the effects of marginal revenues and costs. The methodology is thus disaggregated to the individual company, whereas the command economy could only effect its calculations 'top down'. Each methodology thus exactly matched its political context: freedom with responsibility, or control. Kantorovich's undoubted intellectual and technical success eventually led to social failure in its application, a fate shared by Alfred Nobel, whose father invented the plywood whose manufacture Kantorovich later optimised with his models, and who himself developed the dynamite used in armaments. Nobel was shocked by reading his own premature obituary in a French newspaper saying that 'the merchant of death is dead... Dr. Alfred Nobel, who became rich by finding ways to kill more people faster than ever before, died yesterday', a comment which led him to establish the world famous Nobel Peace prizes as a more positive attribute to his life and work. These are also significant case histories of the interaction of technology and society.

Price is the leading signal in a market economy, preceding and causing profit. If the sales price of the output is above the purchase price of the factors of production, the profit is positive. Price is a complex signal. The three issues needing resolution in a market economy are classically stated as

- the allocative issue: which goods and services are to be produced for consumption?
- the technical issue: how are these goods and services to be produced and distributed?
- the distributive issue: who should have these goods and services?

These issues, known as the 'what, how, and for whom' questions, are all signalled by price. Every single price in a market economy has coded within it and conveys this threefold information signal. Price signals the relative consumer value of any product, indicating to what degree consumers want this product compared to other products. Price is therefore quintessentially relative. At the same time, price incorporates information on how efficiently the product has been produced and delivered. The same product produced in a more efficient way, by a more efficient technology, will have a lower and more competitive price and therefore, all other things being equal, will attract demand away from the first product. Finally, price signals who should have which products, since it contains producer income information. The price of a product whose production has been outsourced to a low wage economy will contain a low income signal, and will in effect allocate less product to the producer than a product made in a high wage economy, whose price will indicate a higher allocation to the producer. Whilst price is a very smart signal in blending three information components in this way, it is also a very difficult and complex one to analyse, since it rarely discloses its composition into these three constituent elements. It is saying three things at once. Is a price high because consumers really get huge benefit from this product, or is it high because it has been made in a very inefficient way, or because it has been made by people who expect high product allocation back to themselves? It may also be that price is high because the market is not competitive, though this can be relatively easily determined by the level of competition in the market concerned.

We therefore need to develop an understanding of how technology operates in a market economy. In preparation for this we first examine the key role of productivity in the economy.

3.2.5.2 Productivity in the real economy

Due to technology, productivity has grown exponentially and had a huge impact on humanity, through its effect on economic well-being and its social and political structures. An ever growing plethora of new products and services characterises consumer society, which in turn characterises humanity. It is a simple arithmetic truth that increased real productivity leads to increased real prosperity, at least when this is measured in terms of material goods and services available per capita. This truth remains even when financial market phenomena appear to obscure it. For example, pensions payable in an advanced technological society depend entirely on current productivity, and not on previous financial savings. Even immense prior financial savings into pension provision will be inadequate if productivity drops. In fact, increased pensions saving will reduce current consumer demand and lead towards unemployment and an output recession in the economy, thus proving counter-productive. Current rules for people to work longer before retiring will have no effect if their longer work period does not increase the total gross domestic product of the economy. By blocking the promotion of younger talented people to senior executive roles, it may well do exactly the opposite, and actually retard economic growth. On the other hand, if productivity were to further continue to grow exponentially, then we could all retire now with adequate provision of all the goods and services we might need or simply want. So whilst money is the modus operandi of a market economy, any meaningful view of the economy must be defined in real terms rather than in financial terms. Politicians are currently mesmerised by the simplicities of financial markets, and often fail to see the real effect of real productivity in the real economy which drives these financial indicators. It is like trying to drive a car by manipulating its speedometer. Productivity is the car's engine, money its lubricating oil, and financial indicators merely part of its instrument dashboard.

Productivity is equally important in considering the causes and remedies of the current (2008–) financial crisis. If gains in productivity are not wholly fed through into real wages, then aggregate demand will be insufficient to purchase the output potential of the supply side. If the gap between output GDP and consumer income is funded with consumer debt, then this will prove un-repayable in future periods, since real wages will remain inadequate. If governments then seek to eliminate financial deficits by cutting the real economy, then employment, output and the standard of living will decline, potentially dramatically.

It makes little sense to allow the financial artefact to dictate to human reality when all real resources are available to generate further output. In this sense, the technical artefact of the financial economy is determining human outcomes. It should not be allowed to do this. There is an alternative: creation of a non-debt-incurring citizen's income.

Imagine a hypothetical state with a high technology, fully automated economy, abundant goods and services and no workers – there would be no wages, and no effective consumer demand to purchase this fully automated output. This is the outcome enthusiastically envisaged as early as 1836 by J A Etzler in his 'The Paradise Within Reach of All Men, without Labour by the Powers of Nature and Machinery'. Output would have to be allocated to the population by means of vouchers or a centrally allocated citizen's income. Failure to do this would lead to unsold or unallocated output and consequent economic decline. Hypothesising this state of full automation demonstrates that productivity can curiously and counter-intuitively lead to economic recession if consumer income and demand is insufficient. This is a set of conditions which can arise jointly.

The current global economy is not yet in this mythical fully automated situation, but as Vaclav Smil points out, the term 'manufacturing' is already anachronistic in that very many consumer products are in fact never touched by human hand in their 'manufacture'. This thought experiment demonstrates that technology can drive productivity towards a fully automated zero-labour position, meaning that some proportion of consumer demand may need voucher, or citizen's income, funding. In his paper 'America's Exhausted Paradigm: Macroeconomic Causes of the Financial Crisis and Great Recession', Thomas Palley advances the argument that US real wages have failed to keep pace with productivity, leading to deficient consumer demand, a gap which has been made up by consumer credit which then becomes un-repayable. Palley's policy solution is to raise real wages. Whilst Palley's analysis is cogent, the argument of this book is that advancing technology makes a reduced wage component of GDP inevitable, so that a citizen's income is the only workable solution. This is more fully set out in my paper 'The 2009 Financial Crisis – A Neo-Keynesian Diagnostic and Policy Response', available at http://tmseu.netgates.co.uk/financialcrisis.html

Technology is therefore mediated to humanity through the economic artefact of the market. Even in western economies, mega technologies such as power generation and transmission, water and sewage, road infrastructure and rail transport were managed by the state. In the second half of the twentieth century there was a virtually universal move away from state enterprise technology, where mega technology appeared to be in the hands of the politicians and therefore as Feenberg declares it 'politically contingent', to being market contingent.

There are few if any models of how technology propagates through a market mechanism, since this would require a rare cooperation between experts in science, and technology, business analysis and philosophy. The commercialisation of technology proceeds according to a model which has three determining criteria which the technology must meet.

- **Firstly** the technology must have a **positive downstream business case**, in other words it must be capable of generating products and services whose delivered value exceeds their delivered cost.
- **Secondly** the technology must demonstrate **competitive price/performance** rating, – it must be demonstrated to achieve the same or greater outcome benefits than other technologies in the same application, at a lower delivered cost.
- **Thirdly**, it must find a route to market – a set of dealers, retailers, specifiers or systems builders who are able to deliver the product or service which incorporates the technology, with this **value chain being viable** economically, that is, profitable.

A good example of this is the smartcard or 'carte à puce' – integrated circuit technology on a credit card – which was invented by a French journalist Roland Moreno, who, between 1974 and 1979, took out 47 patents in 11 countries on his invention. Major companies such as Gemplus were later spawned from this technology. The smartcard was demonstrated in laboratory conditions to have immense data storage and logical processing capability. Applications ranged from identity applications including full secure passport functionality, to medical data including examples of full chest X rays to enable mobile medical consultation, to airline ticketing, to financial payment applications from credit/debit card, to the stored value of an electronic purse. The technology then had to pass the determining criteria of the above model in order to get to market and engage with humanity. This required economic market modelling to ascertain whether in each application, the 'business case' was positive – that consumer demand would be sufficient at an above cost price, that competing technical solutions did not have a better price/performance, and that a profitable route to market existed or would evolve.

In financial payment applications markets, US banks had already migrated all transactions to real time on-line systems, backed up with neural network security checks, and so were less interested in the security a smartcard offered. European banks had different security profiles and so were more eager to adopt the technology. On the other hand, an electronic purse application which had user appeal, had no business case. The cost of installing expensive terminals in pubs and newsagents to allow micro-expenditures on drinks and newspapers proved to be too high, and the transaction processing cost itself exceeded the typical margin on such transactions, certainly as long as governments made cash available at no cost. Two-dimensional bar coded paper slips proved a more cost-effective solution for airline tickets than the smartcard, which therefore tended towards financial payment and loyalty card applications markets. This determining model for the commercialisation of technology can be used to assess strategies for market deployment of a technology by researching and estimating their logic and value in advance, or the model will determine the deployment of a technology in reality.

Another shift in the deployment of technology in a market economy has taken place in supply side structures. The mid-twentieth-century pattern which had emerged was one where state agencies controlled technology development and deployment in most macro industry sectors. Thus, for example, in the UK the Central Electricity Generating Board managed budgets for proprietary technology development, as did British Rail with its ill-fated high speed train.

In 1979 in the UK, the newly elected Thatcher government radically restructured the economy by its thrust of privatisation. Major industries such as power generation and transmission, water, and railways were transferred to the private sector. The assets involved had been constructed over many decades; privatising them forced an attempt to value them. The total railway infrastructure was transferred to the now defunct Railtrack company for a fraction of its historic or replacement cost, since this was the only way to attract investors to a company whose revenues would be limited to the charges for train slots it was able to set. Operating railway franchises went to service operators like the First coach company and Richard Branson's ubiquitous Virgin group, with French rail operators involved in the Connex franchise.

The power generation sector was largely sold to German power generators, later to become Eon and RWE, with the result that new technology plant orders went to Siemens and ABB.

Their advantage derived from the less statist model for technology management that the German government had operated compared to the British. The British model had located all technology development and intellectual property rights in the state operators in the power and rail sectors. The Central Electricity Generating Board and British Rail respectively ran their own technology and R&D centres and owned the technology. This relegated British suppliers such as Parsons to the status of mere manufacturer, without the IPR enabling them to supply markets in other countries. The German model on the other hand contracted technology development to its suppliers, such as Siemens and ABB; the French did the same to Alstom: so these companies were then able to enter, and in fact dominate, the global power- and rail-technology markets.

As in the comparison with the command economy, this experience demonstrates how important structure can be for the development of technology through R&D, and the ownership of IPR, and the consequent competitive positioning towards new market opportunities. The UK gained enormous media attention from its privatisation initiatives, and the continental European countries were often made to look as though their progress towards new private sector models was deficient and retarded. In fact, they approached the transition with a more carefully calibrated strategy and took the technology lead.

A similar differential applied to French management of the public/private sector divide. Instead of transferring water, rail and power infrastructure to the private sector, the French government retained ownership of the assets, and issued contracts to the private sector for five-year management of these assets. This model has also achieved far better outcomes – an integrated standardised high technology railway network, an advanced emission-free nuclear power sector delivering over 80 per cent of the nation's power requirement, and a high technology water infrastructure, incorporating extensive new technologies such as vacuum enclosed waste-water treatment plants and advanced membrane technology. Meanwhile the UK's rail system suffers from the incompatibility of no less than three power systems, ranging through 25KV AC overhead catenary, high speed diesel, and 750V DC third line power, meaning that it is impossible to run one train over the total network.

Its London stations are all also termini which don't allow through traffic. As mentioned above, British Rail's high-speed train failed, whilst tilting-train technology was successfully developed and implemented by ABB, and by Fiat, Italy, with its Pendolino system.

Since the privatisation of these industries, and an important factor in deciding their privatisation, responsibility for technology has shifted in three ways:

i) to the private sector
ii) upstream from user industries to technology dedicated companies
iii) internationally to rationalise and economise from multiple replicated national technology development.

This has had the further result of increasing the concentration ratio of these industry sectors to a situation where a very small number of very large companies supply the technology and its downstream products to all global user industries. Thus boiler and generator technology for powerplants is no longer developed by a national state agency like CEGB, but by Siemens, Alsthom, Toshiba and Westinghouse. This rationalisation does have the benefits of cost efficiency and technology effectiveness, as well as cross fertilisation from analogous technologies developed by the same companies for different sectors, but it does create global monoliths. These could be seen as a new threat to the democratisation of technology, but competition rather than monopoly characterises the global marketplace so far, and this mitigates against abuse. It appears that these private sector multinational corporations are far more vigorous and energetic in developing and deploying technology than their former state-agency counterparts ever were. Even the despised business person is more accessible than the average state bureaucrat.

Due to these phenomena, technology, rather than being politically contingent, is market economy contingent. This is a more dispersed version of democratisation for technology, since every consumer has a choice. Even though that choice may be limited or constrained, it is a direct choice. It can be organised and lobbied to a wider consumer set through Internet and mobile phone networking, a process which has proved feasible and effective.

The technology which ***is*** politically contingent is **military technology** and this does have significant feed-across effects to civilian market technologies. Military technology clearly shapes humanity. As the sword and spear were superseded by the gun, the machine gun, the bomb, the nuclear bomb and now by 'smart' weaponry and potential future satellite weapons, as well as chemical and biological weapons, technology definitely retains the status of an artefact which threatens its creator – humanity. Military technology is often the initiator of generic technologies which then feed into commercial market applications. Companies like Hewlett Packard grew from initial supply of computer technology to the US Korean war effort. Technologies such as night time goggles were pioneered in military applications before being commercialised for hunting and other applications markets.

3.2.6 Society

Society is a network of human beings which brings added value, organisation, and stratification. The network can be defined along any axis. In primitive societies, network membership is defined by the happenstance of birth into a local geographic economic group. Primitive societies are defined by their shared economic needs and activity. Advanced contemporary societies are defined more geo-politically than economically. The definition may be entirely geographic, but in some cases it is also cultural or ethnic. These two definitions can clash, and this clash has been the root cause of much conflict and war. Hindu and Moslem were unable to co-exist in the Indian sub-continent, forcing partition into India and Pakistan. Catholic and Protestant religion and culture struggled to co-exist in Northern Ireland. Serb and Moslem fought bitterly in Kosovo. War can also alter geographic boundaries, so that whole populations can find themselves absorbed into a different nation-state. It is not unusual for a family home in central or eastern Europe to have found itself belonging at separate times to four different nation-states over a period of just over 100 years. In other contemporary cases, geographic nation-states have sought to offer a multicultural society, with legislated religious tolerance, so that the definition of society becomes geographic, with consent to a civic formula.

Membership of these geo-political societies is by citizenship, which in turn can be defined as: a birthright, religious affiliation and historic rights as in the case of Israel, or by formal application procedures for immigration. Citizens are granted passports or other identity documentation, which allows them certain rights of membership of the society – usually those of abode, economic participation in the workforce, and democratic representation, along with social responsibilities – for example, of taxation and observation of the society's legal jurisdiction.

Economic society is often, if not very frequently, organised internationally into multinational corporation employers across several national boundaries, whilst nation-states define society by geographically defined political groups. This multiple definition of society by economy and geo-polity also gives rise to tension, for example when a multinational corporate employer is not uniquely a member of only one geo-political society but transcends many nation-states, in some cases being very much larger in resource than the nation-state itself. At the micro level, society may simply be defined according to shared interest, for example a radio society, or the many interest specific student societies that flourish in universities.

Each individual can thus be a member of several societies, a member of an extended family, a citizen of one nation-state, a worker with a company based in another country, a member of a friendship group, book club etc.

Society exists because it is partially unavoidable, and then continues because it adds value. Individuals are usually born into at least the small society of an extended family. Total individualism is not an option. Then as wider forms of social networking develop and the individual participates in them, a one-way ratchet effect operates, making it extremely difficult for the individual to dissociate. Society is inescapable. In developed societies, people are born into a nation-state and do not have the option of opting out. To achieve any stability, this entrapment is balanced by the benefit of added value, and is retained by inertia. Society enables an economic output which is greater than that which can be achieved by the sum of independent individual action. Society is therefore an implicit trade-off; the constraint of society membership in exchange for its benefits. This is a choice which is rarely stated explicitly, or evaluated.

As society evolves from micro hunter-gatherer societies to still small pastoral and horticultural societies, the productivity of its technology creates a food surplus. This then allows a diversification of labour, the production of other commodities, trade of those commodities for food, and the accumulation of wealth which in turn develops social power structures. In this way, feudalism emerges from agricultural society. Social structures are therefore a result of technology operating through the economy. Any technology requires an organisation of its factors of production. This organisation generates a social structure. Specifically, the organisation requires a stratification, since tasks and roles are less equal than they were in a hunter-gatherer society. Production and consumption are less directly related in the more complex technology, a phenomenon Karl Marx identified as 'alienation'. The reduction or even total loss of this link reinforces the shift to a feudal model where the worker produces for the lord. Military might is a typical new commodity which the lord then applies to enforce feudal society. Might is made to be right as the feudal power seeks to legitimise itself, often through religious claims such as 'the divine right of kings'. Church and state conspire to maintain the system. But this social system suppresses and exploits, in the extreme by enforcing a serf society as prevailed until the nineteenth century in Russia: Tolstoy famously being the first to release his estate's serfs.

The shift from pastoral and horticultural society to feudal society therefore results from

- higher productivity food production technology
- a consequent food surplus
- release and diversification of labour
- production of new commodities
- trade of new commodities and food
- trading led accumulation and concentration of wealth
- de-linking of individual task of production and consumption
- de-linking of individual role as worker and consumer
- emergence of different tasks and roles across individuals in the production process
- emergence of status in production organisation
- emergence of social stratification between workers and between workers and lord
- accretion of military commodities by new feudal lord to reinforce new social structure

Very similar factors operate in transforming a feudal social structure into an industrial society.

The technology of an industrial society needed different working patterns, different skill mixes, different configurations of labour, and a different geographical concentration of production into urban factories. Mass production generated exponential growth in output, which, in turn, required consumers to gather in the same urban societies. Agricultural productivity grew yet further, delivering a constant food surplus and releasing labour for the process of industrialisation and urbanisation. The development of industrial technology was itself fired by the major paradigm shift of all time – the Enlightenment. In the sixteenth century, Copernicus, Galileo and Kepler established the heliocentric configuration and orbits of the planets around the sun. The church resisted, held on to its flat-earth theory, and subjected Galileo to house arrest. But the outbreak of enquiry and discovery was unstoppable, and the primacy of reason, rational argument, and logic inevitable. A whole new way of knowing – a new epistemology – emerged, establishing a science of 'knowing how' rather than simply 'knowing that'. Whilst part of technology could still proceed by harnessing known phenomena without necessarily understanding the underlying explanatory science, now a strong process of 'applied science' became possible to deliberately generate technology as reconfigurations of the explained scientific phenomena. The explanation itself, the scientific theory, allowed a greater proliferation of technology than the application of a single natural phenomenon. So, although technology does not necessarily have to be derived from science, the new scientific rationalism of 'knowing how' generated much technology 'know how' as applied science. Once I know Newtonian mechanics, I can engineer a wide range of applied technologies. The same applies to the biotechnology made possible from the scientific knowledge of enzymes.

The timeline of the Enlightenment is set out in the following table (see table overleaf). Whilst a diverse range of ideas emerged, with some thinkers remaining deists and others arguing for physicalism and atheism, the main import of the Enlightenment was its intellectual methodology of questioning and reasoning. Not only was this applied to nature, science and technology, but also to society. Feudal society – with its immense inequality, its endorsing religion, its military enforcement, its torture and execution – was an inevitable victim of this Enlightenment. Christopher Hill, in his remarkable study 'The English Bible and the Seventeenth Century Revolution', claims that once the Bible became available in English in 1611, the realisation that the text and story condemned unjust kings led to urgent widespread intellectual debate on feudalism's 'divine right of kings', eventually leading to the regicide of 1649. Reasoning was changing social structures. Logic and reason are themselves a technology. The same technology leap which investigated the formulae behind gravity and general mechanics, also questioned the social structure. This process took time, spanned two centuries in its formulation and another two centuries in its application, so that early thinkers did not universally challenge the social structure. Francis Bacon endorsed torture, Descartes allowed vivisection of animals, Isaac Newton sought and obtained a cruel execution of money counterfeiters such as William Chaloner in his role at the Royal Mint.

Nevertheless, the Enlightenment developed a set of core concepts of

- an **anthropocentric** world view – humanity is central
- **physicalism** – only matter exists and all human functions are physical
- **endogenous metaphysics** – humanity generates its own metaphysics – none are exogenous
- **human consciousness** is the sole existential reality
- **reason** is the prime aspect and sole authority within human consciousness and the external world
- **reasonableness** directs ethical judgments, leading to liberal politics and economics

These principles challenged the feudal world. Torture is illogical since the victim can be persuaded to say anything whether true or not. Capital punishment runs the logical risk of irreversibility if later proved wrong. Both torture and capital punishment breach moral reasonableness, although this remains a relative, rather than an absolute, concept.

So technology shifted the social structure from feudalism to capitalism and into democracy, both by generating a huge increase in the output of goods and services, requiring new configurations of production and consumption, and by establishing the same Reason which critiqued feudal society, found it lacking, and urged its transformation. New social classes emerged. The feudal lord was displaced by the merchant adventurer, the landed aristocrat by the industrialist. An increasingly educated population was necessary for more technical artisan tasks, and the working class demanded and won the franchise. Democracy spread, although slowly, and it was only in the twentieth century that the franchise was extended to include women.

The Enlightenment gave us the Age of Reason, the era of Modernity where fact became objective, intellectual doubt and questioning was the

Table 3.5 An enlightenment time line

Enlightenment time line				
17th century		**18th century**		**main themes**
Rene Descartes	1596–1650			self consciousness is the defining point of and for humanity body/mind dualism, so endorsed animal vivisection nominalism leaves space for God also developed Cartesian mathematics, algebra and geometry
Baruch Spinoza	1632–1677			single body/mind substance has endogenous power pantheist and determinist
John Locke	1632–1704			political and economic liberalism, mind is formed by education considered Christianity 'reasonable'
Isaac Newton	1643–1727			mechanism in absolute time and space motion does not require a divine first mover mathematics of gravitational attraction developed the mathematical calculus, optics and alchemy hostile disputes with Leibniz, Hooke, Flamsteed wrote extensively on religion and Bible prosecuted counterfeiters – William Chaloner was cruelly hung
Gottfried Leibniz	1646–1716			virtual 'mondas' exist with endogenous purpose and action that is, not only Newtonian atoms with mechanistic movement rather like 'selfish genes'? also developed the mathematical calculus
Giambattista Vico	1668–1744			myth, cyclical history and Providence are as important as reason
		Voltaire	1694–1778	a deist but religion must be subordinated to reason
		David Hume	1711–1776	there is no separate soul, reason subject to emotion and will rejected the designer argument for creation

Continued

Table 3.5 Continued

Enlightenment time line				
17th century		**18th century**		**main themes**
		Denis Diderot	1713–1784	the universe does not require a divine designer hypothesised evolution of species by natural selection
		Jean-Jacques Rousseau	1712–1778	also a deist but emphasised the primacy of freedom feelings more important than reason, promotes democracy
		Julian Offray de La Mettrie	1709–1751	total physicalism – refuting Descartes body/mind dualism 'the soul is only .... a physical part of the brain' animals have senses, humans just more
		Baron d'Holbach	1723–1789	total physicalism
		Immanuel Kant	1724–1804	emancipation of mankind through an unconditional acceptance of the authority of reason space and time relative to human perception also demonstrated tidal friction drag retarding global spin
		Georg Hegel	1774–1831	many and varied but developed Aristotle's concept of totality

core methodology, and rationality ruled supreme. Modernity proved very effective if measured in terms of the increase in technical infrastructure, in health outcomes such as infant mortality and life expectancy, in productive machinery, in the output of goods and services, and in economic standards of living.

Post-modernity later challenged modernity, objecting to the hegemony of rationality, and insisting on the human option to debunk logic. The feel factor challenged reason. In post-modern society several transformations occurred. Modernity had elevated the notion of function, and had driven functional efficiency ever upwards. Function was therefore more valued, more important than status and image. But post-modernity reversed the emphasis of modernity. Suddenly status became once again more important than function. Image became more important than content, a phenomenon evidenced by the high value placed on product brand, even when the product was technically and functionally undifferentiated from less branded equivalents. In social stratification, the celebrity replaced the scientist, the very rich replaced the intellectual. This post-modern switch to image is also technology dependent, since mass image can only be created and communicated through the mass communication technologies of television, mobile telephony and the Internet. It is, however, an open question of how long image can survive without content, or status without function. Product or personal image can hardly survive a failure to deliver content. The rationality of modernity is based on a deductive logic which is objective and irresistible. So post-modernity will ultimately prove dependent on the modernity it eschews.

The technology-led eras of modernity and post-modernity generated new social definitions. The conscientious educated worker producer was the hero of modernity; the consumer defined post-modernity. The economist Adam Smith who developed concepts of the working of a free market economy, correctly pointed out that the end of all production is consumption. It is therefore foolish to deny the consumer outcome of a productive economy. Production cannot be worthy and consumption somehow unworthy. However it is an equally true corollary that the genesis of all consumption is production. But as technology delivers yet further quantum gains in productivity, the employment base becomes smaller, the complexity of the production process increases dramatically to the point that only the specialist can understand it, and the consumer becomes blissfully unaware of the production process. Outsourcing of production to lower-wage economies makes consumers in rich developed countries even less aware of the method of production responsible for each product and service they consume. Post-modernity's alienation of consumption from production and the consumer society it produces may then prove long lasting. It is the result of technology.

Whilst technology drives social structures, this is not towards a pre-specified outcome. In some geo-political blocks, technology led a process of social change from feudalism to democracy. But in Russia, a very different outcome emerged from similar technology conditions. In the nineteenth century, Russia was a major industrialising power. Its factory at Votkinsky ,where the composer Tchaikovsky's father was General Director, was the largest in Europe, producing railway steam engines and other industrial products. In 1873, Russia was the largest importer of British textile machinery, taking over 35 per cent of British textile machinery exports. So the profile of productivity via industrial technology was similar to other industrialising countries of the period. The social outcome however was very different. The Tsarist regime proved inflexible to the demands for social reform made by the great Russian novelists, Turgenev, Dostoyevsky, Tolstoy, and others. Instead of evolution of the social structure, the Bolshevik revolution of 1917 inaugurated a communist regime which became ever more brutal up to Stalin's death in 1953. It then atrophied until Gorbachev's perestroika and glasnost of the 1980s and the ultimate reform towards a very managed form of social democracy from 1990. The persistence of this statist version of democracy, in effect a combination of weak democracy and strong feudalism, demonstrates that technology affects social structures but does not uniquely determine them.

The same observation is derived from the examples of varying forms of social structure combined with the same high productivity technology in South Korea and Japan compared to Europe and the US. Asian societies are characterised by stricter social structures than European and American society, but the technology/productivity matrix is largely the same. Once again, technology through productivity affects social structure but does not uniquely determine it. This further supports our claim that a systems network model is necessary to analyse the complex multivariate interactions between technology and society.

Anthropologists study the interaction of technology and 'material culture'. Bryan Pfaffenberger in his 1992 paper 'Social Anthropology of Technology'[111] asks 'What is technology? Is technology a human universal? What is the relationship between technological development and cultural evolution?' He rejects the view that 'necessity is the mother of invention', that each specific technology is an inevitable response to an equally specific human need. Since need is culturally defined and not absolute, so diverse technologies might address that need. There is therefore no one-to-one relationship between need and technology. Instead Pfaffenberg points out that every artefact has two dimensions of definition; one functional or instrumental, and the other its symbolic and social meaning or style, presaging and equivalent to the distinction between function and style, content and image that was later to arise between modernity and post-modernity. There is in fact no way to separately measure an artefact's function and style. The huge diversity in style for the same function demonstrates the importance of style in artefacts. Technology is cultural and not simply functional.

Pfaffenberg's answer is to define a 'sociotechnical system' which links techniques, material culture and the social coordination of labour. A sociotechnical system is a coherent combination of these three factors and is therefore stable, resisting 'dissociation'. As an example of a sociotechnical system, Pfaffenberger quotes south Indian temple irrigation, where the temple offered a 'locus of managerial control' which enabled effective irrigation. He writes

> 'The system linked into a cohesive successful system actors such as kings, canal-digging techniques, dams, flowing water, modes of coordinating labour for rice production, agricultural rituals, deities, notions of social rank and authority, conceptions of merit flowing from donations, conceptions of caste relations and occupations, conceptions of socially differentiated space, religious notions of the salutary effect of temples on the fertility of the earth, economic relations (land entitlements), trade, temple architecture, and knowledge of astrological and astronomic tables (used to coordinate agricultural activities). A human sociotechnical system links a fabulous diversity of social and non social actors into a seamless web'.

'Society', he claims, 'is the result of sociotechnical system building'. A sociotechnical system is 'one of the chief means by which humans produce their social world'.

Contrary to the modernist scientific view, ritual plays an important part in the sociotechnical system, by managing the coordination of labour. In Balinese water temples, in South American Piaroa agriculture, in nineteenth century Sri Lankan wheat threshing, ritual played an important role in the social organisation of labour. Such ritual is often silent rather than a cognitive communication of working instructions. This 'nonverbal form of human cognition' is something Pfaffenberger regrets having been lost. Ritual reinforces the sociotechnical system against its detractors and gives it functional stability. Sociotechnical systems enable human creativity include meaning as well as function, and have non-productive elements. However, the varied styles of artefacts in any one sociotechnical system generate power statements and lead to social stratification and a political society. By altering the allocation of power, prestige and wealth, technology generates a social 'drama'. There are gainers and losers from technological change. Pfaffenberg concludes that rather than cognitive science forging technology, in fact, sociotechnical systems develop technology which then generates scientific knowledge.

Society is therefore defined by technology. The question is whether the ontology of society is that of artefact, whether society becomes a 'thing' in itself, with some degree of independence from humanity which is its sole component. Can the totality, as Aristotle claimed, be greater than, or differentiated from, the sum of its parts? Hegel used this concept to justify control of individuals in what became a totalitarian society. It is therefore an issue which has huge implications. If society does have artefact status, we need to be aware of how this happens and consider whether humanity can counteract its own artefact of society.

Émile Durkheim was a founding figure in sociology, establishing the first department of sociology at the university of Bordeaux in 1895. He postulated the existence of 'social facts' which he considered objective in their existence, independent of individuals, and coercive towards humanity. In his 1895 'Rules of the Sociological Method' he defined a social fact as 'every way of acting, fixed or not, capable of exercising on the individual an external constraint; or again, every way of acting which is general throughout a given society, while at the same time existing in its own right independent of its individual manifestations'. He considered suicide a social fact, and in his 1897 study 'Suicide' sought to explain differential rates of suicide between Protestant and Catholic communities by differences in the degree of social control. Crime is also, according to Durkheim, a social fact with an evolutionary social function, as are law and religion. 'God', he famously wrote in his 1912 'The Elementary Forms of Religious Life' 'is society writ large'. This analysis fundamentally saw society as an entity, independent of its individual components, an artefact. Society becomes 'reified' – a 'thing' in its own right.

The French anthropologist Louis Dumont was later to challenge this view, arguing that societies could be either individualistic or holistic, and showing that examples of both types could be found in human history. Nevertheless, Durkheim's analytic formed sociology's version of holism, an important philosophical concept that totality is differentiated from and exceeds the sum of its components. Jan Smuts, the South African lawyer and statesman, in his 1926 'Holism and Evolution' defined holism as 'the tendency in nature to form wholes that are greater than the sum of their parts through creative evolution'. Holism fed into later systems thinking. It is in opposition to reductionism, which on the contrary sees all phenomena as reducible to their smallest component. In reductionism, the 'special' sciences of chemistry and biology reduce to physics, since all higher level systems phenomena are considered explicable by their constituent physics. Not all physicists subscribe to this. David Bohm, who challenged Neils Bohr's Copenhagen interpretation of quantum mechanics, published his 'Wholeness and the Implicit Order' in 1980. Bohm's 'ontological holism' regards the undivided whole as the prime phenomenon, with sub-modules packaged within it. For Bohm, matter and consciousness co-exist in this totality. The recognition of meaning in vision exemplifies the contrast. Reductionist methodology considers that complex patterns, for example the picture of a human face, can be recognised by digital mapping, by a unique combination of pixels on a screen. This methodology however often fails to distinguish the same human face in different expressions, for example, when smiling or frowning. To recognise that the scowling or smiling face is in fact the same person requires 'pattern recognition' technology – a 'top down' interpretation which is different to the 'bottom up' digital methodology. The 'bottom up' concept borrows from reductionism, the pattern recognition methodology from holism.

In summary, technology-led productivity has driven humanity through several conditions, each characterised by a combination of: standard of living, social structure, political structure, and human ontology. The following table presents a summarised hypothesis of how this interaction has worked through history.

Table 3.6 Productivity and social structures: how technology drives productivity which drives the human condition

	Productivity			
	low	modest	high	extremely high
Era	primitive	pre-modern	modernity	post-modernity
Concepts	physical survival	settlement	production content	consumption image
Humanity	hunter-gatherer	farmer	worker-producer	consumer
Context	cave	commune village	town city	cyberspace
Economy		agricultural	industrial	informational
Social structure		feudalism	socialism capitalism	consumerism
Political structure			democracy technocracy	celebrocracy

Low productivity characterises the primitive era when humanity lived in caves or simple structures as a hunter-gatherer, following animal and plant life. Human life in this era was about physical survival. The first technologies then allowed construction of rudimentary fences and farmstead housing, animal husbandry and crop cultivation. This deployment of technology undoubtedly had an immense effect on human life, on what it meant to be a human being. As agricultural technologies developed and were deployed, agricultural productivity increased, leading to sufficient food to sustain more people, with decreased employment of people to produce it. Social structures were then able to develop, since role diversification became possible, as not everyone was needed to produce food. Hence the evolution of feudalism described above, when the landed aristocracy formed. Other technologies for clothing and construction developed, meeting the human need for food, shelter and clothing to increasing standards. The social structures meant that for some this remained subsistence, whilst for others opulence.

And so to industrialisation where mass production and mass distribution delivered its immense proliferation of products to the population. It has been said that the American economy developed from the twin effects of Henry Ford's mass production and the mass distribution of the Sears Roebuck catalogue. People moved to live in cities to staff the industrial factories. Urban build and services technologies enabled city life. In this new configuration of people, infrastructure and role, feudalism was replaced by democracy. The emphasis was on production as industrialisation harnessed the logic of the Enlightenment. Education and know-how developed content knowledge. As the white heat of the technological revolution raced onwards, technocracy even appeared to threaten democracy. Productivity did not plateau but continued its exponential growth. The reduction in the working week that this allowed, together with the massive increase in real income for the worker turned that worker primarily into consumer. The consumer society had arrived, with its abundant output allowing choice rather than scarcity, so that the consumer rather than the producer assumed sovereignty. Dire Malthusian predictions were avoided as populations grew together with living standards, rather than the two being inversely related. In the post-modern age, an easy affluence generated by this technology-led productivity, allowed logic to be suppressed in favour of the feel factor, content to be challenged by image, function by status, and community by individualism. It is beyond doubt that what it felt like to be a human being – the human experience, human social and political structures, and the very ontology of humanity – had changed immensely, and that this change was entirely due to technology working through productivity. Humanity is redefined as techno-humanity, and this has immense philosophical import.

With the benefit of these diversions into the theories of sociology, anthropology and physics, can we conclude whether society has the status of an independent artefact ? Human experience and human mythology abound with examples where this at least seems to be the case. We create society which then constrains us. Law is a prime example. The Biblical myth records the story of how Darius, king of Persia, was tricked into passing a law that anyone who refused to worship him alone should be thrown into a den of lions. He had not reckoned on the fact that his Jewish monotheist protégé Daniel would not be able to worship Darius, and so, since the law of the Medes and Persians was considered immutable, was forced by the institution of the law to throw Daniel to the lions. Miraculously an angel intervened and Daniel was spared, but this event has proved incapable of repetition as a scientific experiment! The main method by which society becomes an artefact is the process of institutionalisation. Institutions such as the law – or various protocols, rituals, religions and their priests, academia – are all initially formed to serve a purpose of the time. Human life then develops and in reality requires fresh evolving institutions. But whilst humanity and human life is organic, the institutions humans create are inorganic. Rather than evolving dynamically to meet new requirements, they ossify and can even act contrarily to later human needs and preferences. Institutions need constant dynamic renewal, which they rarely get.

There are many cases where society through its institutions delivers an outcome that no one individual and maybe even no definable group of individuals wants. Involuntary unemployment analysed by Keynes, is one of these phenomena. It really is involuntary. No one wants it, but society through its market artefact delivers it, and humanity seems powerless to counteract the effect of its own institution. Complex behavioural patterns of 'double entendre' can also lead through game theoretic analysis, to outcomes no one wants, due to a lack of knowledge of what individuals actually want.

3.2.7 Ecology

The final interaction in the system network is from redefined technohumanity back to nature. The exploitation of minerals, the levelling of hills and valleys, the creation of huge dams for hydro-electric projects, the building of cities, the concreting and tarmacadaming of vast surfaces, the emission of pollutants to land, sea and air, the extinction of animal and plant species, are all areas where technologically empowered humanity has an immense effect on nature, which has in Heidegger's terms been redefined as a mere resource with no inherent identity and value. Thomas P Hughes in his 2004 'Human Built World'[112] quotes Goethe's, Faustus who in negotiation with Mephistopheles, chooses a land reclamation scheme, expressing his urge to control land and people, and loses his soul. For native American Indians, Hughes says, 'nature was bountifully life supporting' but for the new European settlers it was 'a wilderness to be conquered'. 'Technology' writes Hughes, 'remained a creative tool, but one that stunted and took lives, and dominated and despoiled the environment.' He gives the example of copper mining in Butte, Montana where the Anaconda Copper Mining Company grew to become the fifth largest US company in 1912. Despite some efforts at land reinstatement, Hughes says that 'even today the Butte region remains one of the most devastated in the nation'. Changing the humanity-to-nature interaction from despoliation to preservation, restoration and holistic respect for nature requires a change in the American religious mentality: according to the Bible's book of Genesis, humans are charged to subdue nature. Hughes quotes Lynn White writing 'We will continue to have a worsening ecological crisis until we reject the Judeo-Christian domination axiom'. Hughes, like Feenberg, calls for the creative deployment of ecotechnology to replace historic concepts of exploitation which have led to despoilation.

4
Resolving and Managing the Model

4.1 The behavioural economics view of technology – innovation studies

Whilst there is only a small corpus of literature on the philosophy of technology, there is a wide body of research and published material on the economic history and industrial management of technology. Curiously, this literature is distinct from the philosophy of technology, using the terms 'invention' and 'innovation' rather than the word 'technology' itself. There appears to be virtually no overlap in contributors or contributions between the two fields of study. The Science Policy Research Institute (SPRU) at the University of Sussex, UK, founded in 1965, is a central contributor to the discipline of innovation studies, and a large international network of similar institutes has subsequently developed which is set out on the SPRU web site. Christopher Freeman's 1974 book 'The Economics of Industrial Innovation' and the 1988 'Technical Change and Economic Theory' edited by Dosi, Freeman, Nelson, Silverberg and Soete[113] came from the work at SPRU. It is the work of this network which has informed and guided public policy in technology management, though this also has gone under the title of innovation policy in the creation of 'national innovation systems'.

The core concepts are historical, economic, managerial and political. They are not philosophical, which accounts for the separateness of the philosophy of technology, but suggests scope for convergence between the disciplines, particularly since the work of SPRU and other technology institutes is essentially multi-disciplinary. Innovation studies derives from the work of Joseph Schumpeter who defined innovation as 'new combinations of existing resources', a definition close to the definition of technology adopted in this book. Schumpeter emphasised the tendency for innovation to cluster in defined time periods, and indeed technology development has historically also clustered geographically, either in the early industrial revolution of the UK, or the IT seedbed of Silicon Valley. Schumpeter tried to analyse such time clusters of innovation into long term business cycles. He defined five types of innovation, ie

- new products
- new production methods
- new sources of supply
- new markets, and
- new business organisation

All of these contributed to economic growth.

The historical narrative perspective of innovation studies identifies various ages of technology.

- 1780 to 1840 is the industrial revolution with its factory system, the mechanisation of the textiles industry, and the building of canals
- 1840 to 1890 is the age of steam and railways
- 1890 to 1940 the age of electricity and steel
- 1940 to 1990 the age of mass production leading to
- the age of information technology from 1990.

This perspective tends to obscure the more incremental nature of technology development, both in the emergence of major technologies, but also in their combination with other technologies and in their constant but gradualist improvement.

More analytical content is derived from a rather different historical analysis of four quantum leaps in technology in industrial application.

- The first of these is the **UK** industrial revolution
- the second the **German and US** surge in industrial development
- the third the **Japanese** success in industrial productivity, and
- the fourth the **US** growth in the electronics sector.

The **British industrial revolution** developed and deployed technology with little social structural support. Entrepreneurs invented machines and craft workers worked them. According to William Lazonick,[114] firms like Platt Brothers who dominated textile machinery manufacturing for the UK and export markets, had no company R&D activity. Nevertheless, by 1850, British productivity and income was 50 per cent higher than in other countries. Technology, however, was deployed and disseminated very slowly. It was 50 years before Robert's automated spinning machine accounted for a majority of UK cotton output.

Family firms suffered when second and third generation family owners could not maintain the founder's initiative. US industrial development crucially overcame this problem by separating ownership from managerial control in its 'managerial revolution', with early IPOs onto the stock exchange and the emergence of the corporate executive, generated from the Harvard precursor to the MBA course initiated in 1908. Technology was brought to market and the technology market strategy matrix was managed.

Meanwhile, **Germany and the US** developed the in-house R&D department. Technical high schools were developed in Germany. Christopher Freeman[115] quotes Eric Hobsbawn to the effect that by 1913 Germany was training 3,000 graduate engineers a year compared to Britain's 350. By 1914, the German electrical firms AEG and Siemens each employed 50,000 staff, whereas no British electrical firm employed more that 10,000. General Electric and Westinghouse were the huge US comparable companies. Together with government research institutes, this was an early German example of a national innovation system, inspiring and enabling a technocratic culture. The cultural difference remained, and late in the twentieth century, being a Dip Ing in Germany was a source of pride and social location, whilst for British graduates, an education in the classics was seen as the route to social status. Britain proved capable of invention, but lagged in innovation. It proved deficient in the new electrical and chemical growth sectors, whereby Germany and the US displaced Britain as the leading industrial power. Britain's focus on its empire diluted its potential in the new modernity.

Lazonick analyses **the 1970s and 1980s Japanese industrial boom** as due to

i) cross shareholding, giving stable ownership whereby leading companies by 1975 owned 60 per cent of each others' shares which were then not traded
ii) extensive bank debt with gearing ratios up to 7:1 supported by the Bank of Japan

iii) the stability of lifetime employment, making investment in staff worthwhile.

Efficiency was the key Japanese success factor; Lazonick writes that in the 1980s, Japanese production of DRAMs (direct random access memory chips) was 40 per cent more efficient than US production, and that by the early 1990s, Japanese factories deployed over seven times as many robots as US manufacturing plants, raising manufacturing quality and lowering cost. Organisation of industrial activity was the key technology in the Japanese success.

The later US Silicon Valley phenomenon derived from a free wheeling entrepreneurial technology culture, supported by risk taking venture capital and stock based executive compensation. Technology-led business development depended on the nature of the firm, which Lazonick points out is better seen as the 'innovating firm' creating new technology market strategies, than the 'optimising firm' of classical economic theory which takes technology and market price as fixed and given.

In current industrial technology development, networks and consortia are important drivers, as innovation tends towards the model of 'open innovation' triumphed by Henry Chesbrough and others in his book 'Open Innovation'.[116] In biotechnology research, the public sector Human Genome Project was a wide consortium, as was the rival private sector Celera company genome project. Inter-company R&D is now commonplace in many industries, Nokia and Microsoft being a recent example. Interaction is key to innovation, and exclusive private development is less creative, a factor which inhibits more effective industrial technology development in Russia, where all-Russian 'national champion' initiatives are forever preferred, despite their lack of deliverable result.

Innovation studies has its root in economics. Jan Fagerberg[117] identifies the 'Marx-Schumpeter model of technological competition'. Marx held that firms will innovate to gain competitive advantage which will then be signalled in the market economy through higher profit, attracting and indeed forcing other firms who want to survive to also adopt the new technology.

But innovation studies also seeks to explain innovation. According to innovation theory, innovation, or technology, is contingent. It depends on social structures, the nature of the industrial company's organisation and inspiration, on wider public institutions, and on public policy.

Policies can therefore be effectively implemented to improve innovation by acting on these causal factors. Key among these factors are

- enabling and encouraging firms to behave innovatively
- creating and fostering networks of innovation
- harnessing university contribution to innovation in industrial partnership
- the availability of venture capital finance
- the existence of protective IPR
- the nurturing of geographical clusters of innovation
- understanding 'path dependency' of technology, and avoiding its constraint
- the understanding of the global nature of innovation

As an example, one 1992 study of R&D investment in Japan found that increasing IPR patent cover by three years would increase R&D investment by firms by between 3 per cent and 8 per cent, whereas reducing it to zero would reduce R&D investment by 40 per cent in the electrical sector and 60 per cent in the chemicals sector.

Innovation studies then conducts a series of such microeconomic studies and feeds the results into government policies to enhance innovation in the economy. Its more recent theory has been 'endogenous growth theory' initiated by Romer, in which R&D investment is determined by businesses calculating the cost and benefit profiles of technology.

In some ways, innovation studies refers to the network model of technology developed in this book, in that science is included in its study and policy advice, and the productive economy is the result of technology. It does not however, analyse nature, science, technology, economy, society interaction thoroughly, and does not consider humanity in symbiosis with technology from a philosophical perspective. It does however focus on the effect of society on technology. Its emphasis is on how the social structures of and around the firm, society more generally, and its political structures and policies affect innovation. If it has a philosophy of technology, it is an implicit one, and would probably be that technology is socially determined, but in this particular behavioural way via the industrial firm and its social context. Technology transfer, which is encouraged by the open innovation model, foresees technology developed by one social culture being transplanted to a culture that did not generate it, may possibly never have developed it, and may indeed prove unable to accept it being grafted on.

4.2 Models of the management of technology in a market economy

We have pointed out that technology is market contingent. The initiative and responsibility for technology lies with the private sector of the market economy. Even where technology is politically contingent, governments often contract private sector companies to undertake technology research and development, usually known as R&D. This section sets out the extent and a type analysis of technology management in the global market economy, and explores some paradigms of how private firms manage technology in society.

The UK government publishes an annual R&D Scoreboard in which it reports and analyses both global and UK private sector R&D. The 2010 scoreboard shows that in 2009 the R&D investment by the leading 1,000 companies worldwide amounted to £344 billion. Of this, 82 per cent was undertaken by just six countries, namely the US, Japan, Germany, France, Switzerland and the UK. Of this global R&D investment, 52 per cent was in just three industry sectors: pharmaceuticals and biotechnology, technological hardware and equipment, and automotive. Like most phenomena in a market economy, technology has thus become very concentrated in six countries and three sectors.

The following figures taken from the UK R&D Scoreboard show

1. The concentration of global private sector R&D in the US and Japan and four European countries

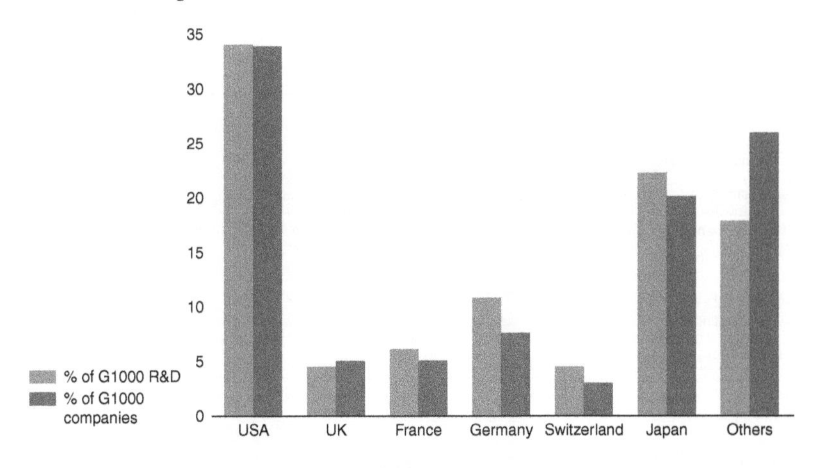

Figure 4.1 Distribution R&D expenditure by country

2. The focus of global private sector R&D in three major industry sectors

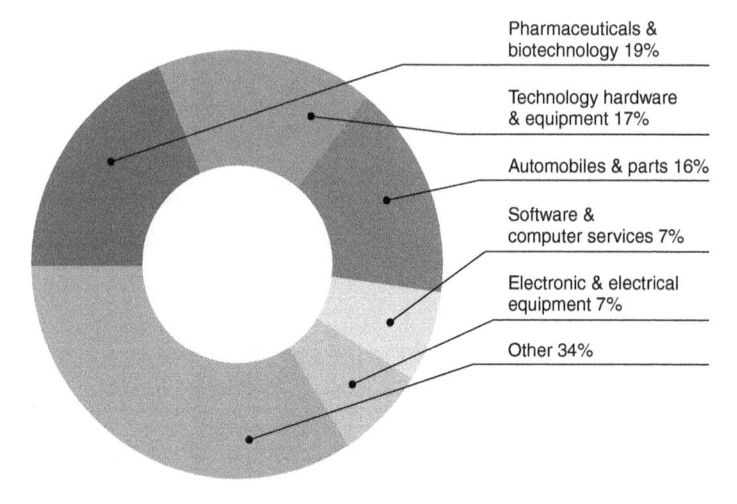

Figure 4.2 Distribution of G1000 R&D expenditure

3. The sector specialisation of the leading country R&D investors, showing how Switzerland and the UK focus on pharmaceuticals and biotechnology R&D, whilst South Korea focuses on electronic and electrical R&D, Germany on automotive R&D, and the US maintains a more balanced portfolio of R&D across these sectors.

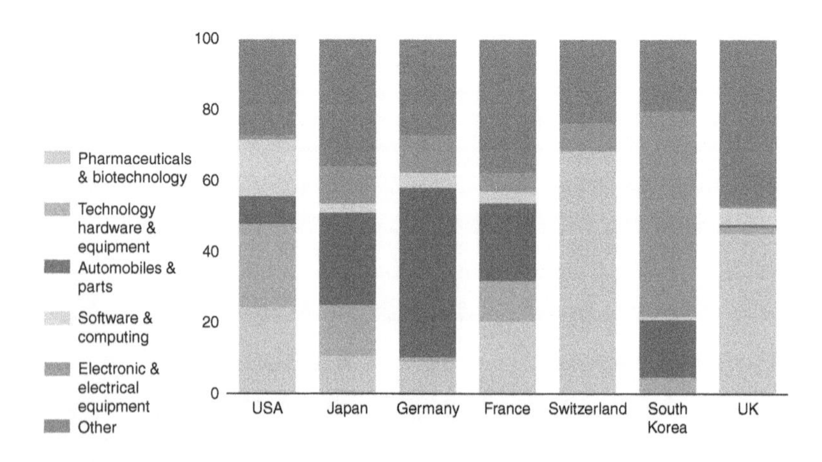

Figure 4.3 R&D expenditure by sector and country

4. The leading companies responsible for global private sector R&D investment

Table 4.1 Leading companies in global R&D

Rank 2010	Company	Sector	Country	R&D (£m)	Growth in R&D over last year (%)	Rank 2009
1	Toyota Motor #	Automobiles & parts	Japan	6,014	–6	1
2	Roche, Switzerland	Pharmaceuticals & biotechnology	Switzerland	5,688	9	4
3	Microsoft #	Software & computer services	USA	5,396	–3	2
4	Volkswagen	Automobiles & parts	Germany	5,144	–2	3
5	Pfizer &	Pharmaceuticals & biotechnology	USA	4,802	–2	6
6	Novartis	Pharmaceuticals & biotechnology	Switzerland	4,581	2	10
7	Nokia	Technology hardware & equipment	Finland	4,440	–6	8
8	Johnson & Johnson #	Pharmaceuticals, & biotechnology	USA	4,326	–8	7
9	Sanofi-Aventis	Pharmaceuticals & biotechnology	France	4,060	0	12
10	Samsung Electronics #	Electronic & electrical equipment	South Korea	4,007	8	18
11	Siemens	Electronic & electrical equipment	Germany	3,805	2	20
12	General Motors USA #	Automobiles & parts	USA	3,758	–24	5
13	Honda Motor #	Automobiles & parts	Japan	3,746	–4	11
14	Daimler	Automobiles & parts	Germany	3,700	–6	13
15	GlaxoSmithKline	Pharmaceuticals & biotechnology	UK	3,629	10	21

Continued

Table 4.1 Continued

Rank 2010	Company	Sector	Country	R&D (£m)	Growth in R&D over last year (%)	Rank 2009
16	Merck #	Pharmaceuticals & biotechnology	USA	3,619	22	25
17	Intel #	Technology hardware & equipment	USA	3,501	–1	17
18	Panasonic	Leisure goods	Japan	3,445	–7	14
19	Sony #	Leisure goods	Japan	3,308	–4	16
20	Cisco Systems #	Technology hardware & equipment	USA	3,225	1	22
21	Robert Bosch	Automobiles & parts	Germany	3,179	–9	19
22	IBM #	Software & computer services	USA	3,061	–10	15
23	Ford Motor #	Automobiles & Parts	USA	3,034	–33	9
24	Nissan Motor #	Automobiles & parts	Japan	3,030	0	23
25	Takeda Pharmaceutical#	Pharmaceuticals & biotechnology	Japan	3,014	64	n/a

accounts not prepared using IFRS

Despite technology's key role as an economic driver, both the economics and the philosophy of technology receive little attention in academia. Equally, processes for the management of technology are not well developed and deployed in productive industry. Through its effect on productivity, technology is fundamental to prosperity, and is therefore of immense social interest. At the same time it is fundamental to business profitability, through cost reduction efficiencies, new product market opportunities, and greater product functionality, all of which enhance a company's competitive market positioning, and hence its sales potential and profit margins. Business technology management processes therefore need to be better defined, understood, and implemented. Governments who fund commercial R&D with public funds, and banks who fund it with private sector funding, also need to understand the technology management process more clearly.

The **key objective of technology** for a business operating in the global competitive market economy is to **maximise competitive price/performance** positions. In effect this means seeking to minimise the actual ratio of price/performance itself. To win competitive market share, a product or service has to offer the maximum performance benefit for the minimum consumer price. Maximum performance includes the very existence of the product or service if it is new to the market, and all the defining functionality it incorporates and evolves through its subsequent design generations. Its availability to the consumer through efficient distribution channels and the availability of after sales service are part of this performance measure, which is the sum total of all the consumer benefit the product or service offers. Incorporating additional functionality from ever-evolving technology into attractive customer-focussed designs is essential to keep the performance measure competitive. In post-modern consumer markets, image is as important as content, so that culturally approved cool product design is as important as products' operational functionality, sometimes even more so, as amply demonstrated by Apple's laptop computers and mobile telephones. At a basic level, management of technology should always be targeting (1) competitive product performance, through optimum functionality from leading edge technology, as long as a business case can be established: that is, where customer demand is positive at the proposed price, and (2) production and distribution costs, so that optimal functionality and design is delivered to the customer at a competitive price.

For the optimal management of technology, technology needs to be defined by type and by source. Types of technology include enabling technology, for example CADCAM in manufacturing design and architecture; generic technology, for example propulsion and power technologies; application specific technology, for example mobile phone technology; engineering technology in the use of manufacturing equipment and the configuration of labour skills; and product technology, for example lenses for spectacles. Sources of technology are more varied than is commonly considered or exploited. These can include a company's own internal R&D, IPR and patent portfolio, competitors and collaborators with whom it can share technology, universities, contract R&D companies, or commercial suppliers of technology, either incorporated into capital equipment, or available through commercial technology licensing.

The following diagram sets out the complex of options for sourcing technology along a company's value chain.

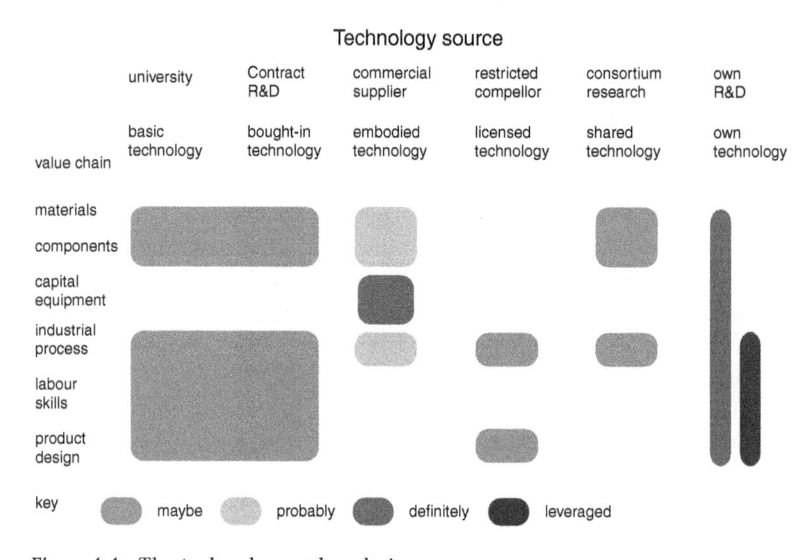

Figure 4.4 The technology value chain

Optimising this option set is extremely challenging, and yet it is core to business success and to the deployment of technology to society through the main channel of the market economy. R&D ranges from 'blue sky' research where scientists and technologists pursue the science and technology per se without being driven by any business objectives, to highly constrained R&D to solve an applications-engineering problem. Companies face difficult trade-off decisions in the allocation of R&D investment, as the outcome is described by a probability distribution populated with low information-quantity and -quality. Many companies implicitly acquire technology incorporated into capital equipment and software systems from commercial suppliers. Numerically controlled machine tools, enterprise software systems such as SAP's R/4, are leading examples of value-chain sourcing of technology. The same applies to materials and components which are bought from upstream suppliers.

In command economies, integration tended to be vertical, along the value chain of each sector. Factories commonly operated their own foundry for example. This was inefficient, since consolidated experience was not used to develop casting technology further. Suppliers also became lazy, unincentivised by the lack of competition to supply captive customers in the same business. In a market economy, concentration, which regularly becomes high with 70 per cent, three firm market shares common in many sectors, is nevertheless typified by horizontal integration. The advantage, compared to the vertical integration of a command economy, is that a more extensive but focussed R&D and investment resource is addressed to technology at each specific point along the value chain, for example casting. As a result, the value chain itself is more fragmented by corporate ownership; typically companies purchase, in this example, finished castings produced by an advanced technology specialist castings sector. A wide range of materials, from ferrous and non-ferrous alloys to polymers and composites, is procured from specialist suppliers who invest to develop their material technology IPR.

There are also many occasions where technology can be licensed by competitors. One example is the Korean piano industry, where Kawai licensed vacuum plate technology developed by Yamaha, Japan, and alongside this agreed some limits to geographical market competition. Contract R&D companies operate laboratories to conduct contract research, and so offer another source of technology in the market economy. The engagement of industry with academia has often proved difficult to manage effectively as a source of technology for the market economy. Nathan Rosenberg in his 'Technology and the Wealth of Nations'[118] documents the effective interaction of academia, consulting companies and industry in the development and global spread of the US chemical engineering sector. The Massachusetts Institute of Technology, the consulting company

Arthur D Little, and Exxon cooperated effectively in this process. Other governments attempted to foster such collaboration, for example, by funding science parks close to major universities, as in Cambridge, UK. This can succeed, but has to overcome tensions from the differences between commercial and academic objectives, drives, and interests. Academics can seem naive and other-worldly to business executives, who in turn seem simplistic in their commercial objectives to academics.

However there is a potential threat from the process of acquiring technology upstream. Current trends in the globalisation of the market economy do open new opportunities for technology strategy, but also create new risks. The opportunity is that increased external sourcing of technology appears a cost effective way of offering leading edge products into global markets. Whole economies, especially in south-east Asia, have benefited from such technology transfer and licensing. Generic technologies are available from market supply sources. Industry/university technology collaboration is actively promoted by government schemes. But the corresponding threat is that of the zero sum game. As each company faces a short-term micro incentive to buy in technology, or to recruit to buy in trained labour, thus saving on its own R&D investment and human training costs, so the spread of such behaviour across the whole economy threatens to lead to under-investment in technology development overall. Partial analyses which show technology migrating typically up the value chain of service operations and manufacturing industry, must ask where short term responsibility for technology ultimately resides in today's global competitive markets.

In our model of the artefacts of technology, market and society, we have considered whether and to what extent the technology artefact drives the market artefact through its productivity effect. We can equally ask the reverse question of whether, how and to what extent, market drives technology. We have explored this by showing that technology is market contingent by having to work through the market algorithm of a positive downstream business case, competitive price/performance positioning, and a viable value chain. Another way in which the market can impact technology is through the development of a market for technology itself, rather than the downstream markets for the products and services developed from the technology.

In a world where focus on core activities is a competitive necessity to corporate survival and success, the inherent opposite tendency of technology development to lead towards diversified outcomes – ever more so as 'basic' research is in advance of and detached from product development – creates strong a priori conditions for a technology transfer market. The infancy of this global technology-transfer market inhibits investment in technology development, by amplifying the risk of redundancy of outcome to the R&D process. Such redundancy easily arises when commercial business management strategies and scenarios evolve in shorter time circuits than successful technology development, or when company 'champions' of a technology project move job within and between companies. A technology-transfer market with extensive coverage and a mature set of trading rules, which provides risk cover against technology-outcome redundancy, has yet to emerge.

Such a market can be shown operationally as

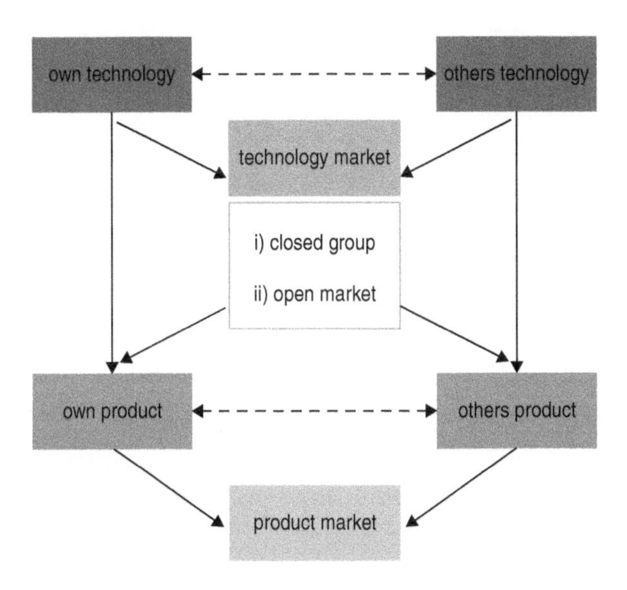

Figure 4.5 Technology market interactions

Companies operating within the technology market will then be able to determine their technology and product strategy according to a simple management process ie

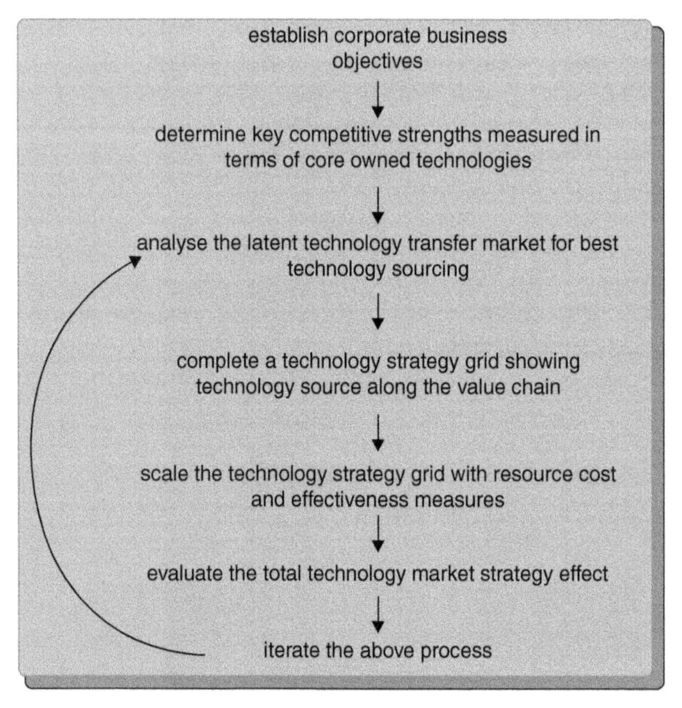

Figure 4.6 Technology management business process

So far we have considered how a company can manage its technology strategy in the external markets up and down the value chain from its own position. Another core question is how private sector companies can report and audit their technology position internally.

Financial reports are well established mechanisms in corporate management, ranging from legal audit requirements to internal group company reporting structures. Whilst these financial reports can indicate future product market business opportunities, they inevitably report the results of past performance, rather than analyse the factors which will generate future business performance. The balance sheet does not value the company's technology holding, its IPR, in a meaningful way. And yet it is the corporate technology portfolio which clearly represents the future business potential, which by being leveraged optimally into product or service market opportunities is the key to profitable organic growth. The question is how this important driver can be best reported and managed. A potential answer is through the incorporation of a formal technology market audit into the business process.

An example of such a process for technology market audits might be

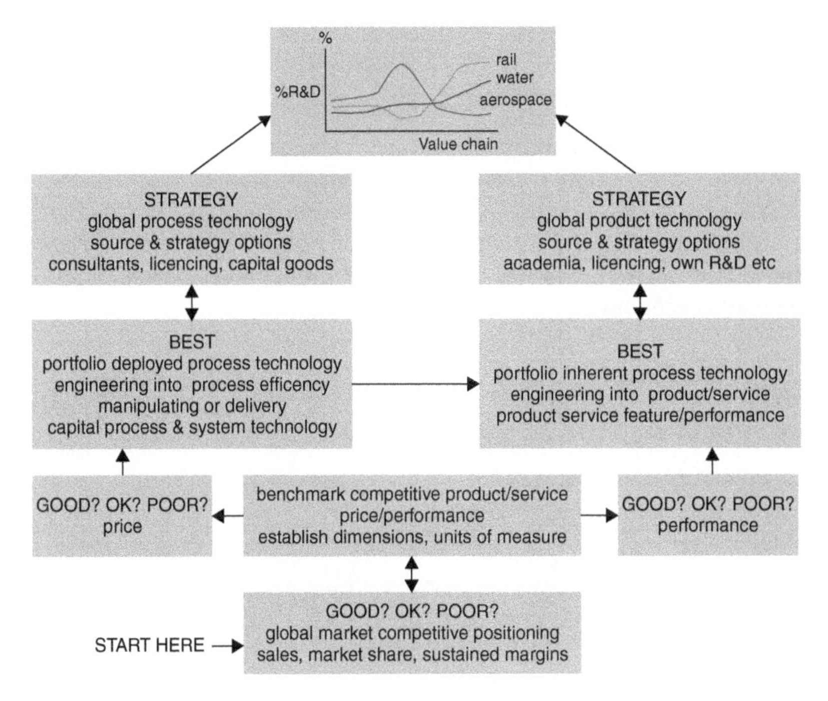

Figure 4.7 Technology business audit process

The diagram is read 'bottom up'

The process proposed is

- Known measures of the company's competitive market positioning are explained in terms of product price/performance which is itself derived from
  - price resulting from the set of process technologies employed
  - performance resulting from technology engineered into the product

- This then triggers a review and evaluation of the company's deployed technology portfolio
- A diagnostic of technology in the industry is developed to identify major trends characterising the dynamic deployment of technology along the value chain
- Management of technology strategy, covering issues of R&D spend and technology sourcing options, is then determined from the competitive product market requirement against this diagnostic of technology in the global industry sector

Just as accountants established accurate processes for reporting a company's past performance, the 'rear view mirror' approach to driving, so business-process analysis needs to develop meaningful technology strategy audit processes which can be widely deployed as more meaningful front-view approaches to managing companies in a technology-led market economy.

Technology has a pivotal role in the process of economic development. Through its effect on productivity, technology has changed the course of economics to allow the population subsistence theories of Malthus and the capitalism disaster theories of Marx to be avoided.

Technology and 'know-how' is embodied in human education, in communications and transportation infrastructure, in physical capital equipment and software, in industrial processes, and in new product and service creation and functionality. These channels for the implementation of technology are the levers for advancing standards of living.

4.3 The model's interactions

So far we have discussed the model's assumptions and its entities. We now move to explore the interactions between these entities which make the model a dynamic symbiosis. To some extent, the discussion of the nature of each entity has involved some development of the interactions with other entities, since it is impossible to define each entity absolutely without reference to its relativity to the rest of the model.

Our initial assumption is that nature determines science. Science is knowledge about nature and so cannot exist independently of nature. It is a statement about nature. However this it not a total statement of the ontology of science. Science also depends on humanity, since human effort is required to investigate nature in order to state science.

And science is dependent on technology, since the microscope and telescope of optical technology have been essential to the progress of science. It also depends on productivity since the rate at which experimental data can be gathered and tested determines the rate of progress of science. Science also depends on the economy, since financial investment and resource allocation are necessary for scientific research funding. Science also depends on the market artefact of the economy, since major commercial companies make decisions to fund science based on their current and future market performance and financial health. Science depends on society, since social recognition of the scientist is a factor in attracting intellectuals to work in science. Governments can choose whether to fund science or not, and which science to prioritise. We therefore see that science is in effect dependent on all of the other entities in the model, the real entities of nature and humanity, and the artefacts of technology, economy and society.

The same analysis can be applied to each of the other entities in the model. Technology depends on nature, since it is natural materials and processes which technology reconfigures. It depends on science, if we define science to include 'knowing that' as well as 'knowing how'. Similarly to science, technology depends on humanity for intention, initiative, and work; on the economy for resource; and on society for status and structure. The economy in turn is dependent on nature for its materials, on science and technology for its production function, on humanity for its labour, and on society for the organisation of production, distribution and consumption. Society is determined by technology and therefore, at least indirectly, on science, productivity and the economy. Society is dependent on nature, even for its geographical location.

If each artefact in the model is so extensively interdependent, what of the two real entities of nature and humanity? Nature, rather than being independent and objective, is in fact subject to humanity, science, technology, productivity, economy and society. If this were not so, there would be no ecological concern. Humanity, through its economic society and the power of its technology, exploits nature for its resources. This is Heidegger's concept which renders nature a 'standing reserve'. The economic production system creates emissions to land, water and air, polluting ground with heavy metals, water with oestrogen containing detergent, and the atmosphere with SOx, NOx and CO_2.

Humanity is dependent on nature. In the original position and first iteration of the model, humanity finds nature 'red in tooth and claw' to quote Tennyson, and the result is a human life which is 'nasty, brutish and short', according to Thomas Hobbes. In the northern hemisphere, it is doubtful that human life could exist at all without the intermediation of technology. Naked humanity needs a Garden of Eden where provision is bountiful and free. This of course makes humanity extremely dependent on technology.

We can set out the extent of the complexity of interdependency in the model in the table shown on the facing page.

Every entity therefore depends on and in turn impacts every other entity. The network dependency is shown diagrammatically as

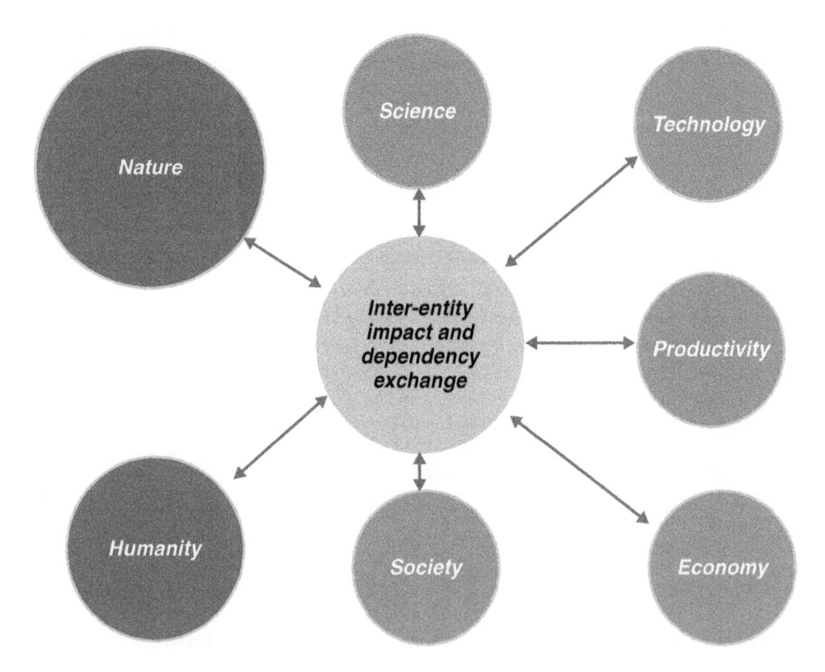

Figure 4.8 Dependencies in the model

Table 4.2 Dependencies in the model

Entity	Dependency	Nature of dependency
Nature	Nature	Nature is an endogenous interactive system, dependent on itself. The water cycle is a good example; ocean water evaporates to form clouds which are driven by wind to precipitate over forests to water crops and animals
	Humanity	Humanity impacts nature through the role of its cognitive intentionality in the exploitation of raw materials, its rearrangement of nature in damming rivers or planting crops, and the emissions of its industrial processes to nature
	Science	Knowledge about nature is a factor enabling humanity's impact on nature
	Technology	Technology impacts nature by reconfiguring nature's materials and processes, and introducing non-natural materials like plastics into nature
	Productivity	Productivity determines the rate over time at which humanity, science, and technology impact nature
	Economy	The scope of economic activity, in investment, production and consumption determines the extent of the impact of humanity, science, and technology on nature
	Society	Social structures including government action determine the above impacts on nature. Society's location decisions impact nature differentially. Social attitudes affect the philosophy of society towards its impact on nature
Science	Nature	Science is knowledge about nature, both 'knowing that' and 'knowing how'. Science therefore fundamentally depends on nature
	Humanity	Science is knowledge by humans about nature and depends on human intentionality, interest, initiative, and work
	Science	Science is also an endogenous system; each element of science potentially impacts all other science. Explanatory models realised from one natural phenomenon often find more generic application

Continued

Table 4.2 Continued

Entity	Dependency	Nature of dependency
	Technology	Science depends on the technology it has itself initially created. The microscope is the prime example of this where optical know-how enabled observation of the atomic world. Mining technology aids the science of geology
	Productivity	The rate of scientific progress depends on productivity
	Economy	The economy funds science, often dependent on market expectations
	Society	Social attitudes impact science, whether elevating the status and power of science and scientists (viz C P Snow's Corridors of Power) or relegating science to commodity status as in post-modernity and celebrocracy
Technology	Nature	Technology is the reconfiguration of natural materials and processes, and so fundamentally derives from nature
	Humanity	There would be no technology without humanity since technology is dependent on human intentionality and action
	Science	If science is defined both as 'knowing that' as well as 'knowing how', then technology is entirely dependent on science. Only if science is defined restrictively as scientific theory ('knowing how') can technology appear independent of science through simple observational 'knowing that'. The science of geology aids mining technology
	Technology	Technology is highly endogenously self interdependent. Constantly one technology enables and supports another, and combines with another in every simple product. Very few technologies are standalone. Schumpeter's 'combination' and 'clustering' are core dynamics of technology.
	Productivity	The rate of development of technology depends on productivity
	Economy	The needs and the outcome results of the economy are a powerful driver on technology
	Society	Social structures, status and attitudes, and government policies affect technology

Table 4.2 Continued

Entity	Dependency	Nature of dependency
Economy	Nature	Economic production exploits nature as resource, both for materials and processes and as a waste repository
	Humanity	Humanity as investor, producer and consumer develops the economy
	Science	Science is itself a sector of any economy, part of the GDP.
	Technology	Technology drives productivity and therefore standards of living
	Productivity	Productivity, that is, the rate of output over time and drives the standard of living. Curiously, if not directly linked to real wages, productivity can exceed consumer demand and so generate economic recession
	Economy	The economy is interdependent through its various sectors and through the dependencies of investment, output and consumption
	Society	Social structures, whether feudal, capitalist, or communist, have great impact on the economy
Society	Nature	
	Humanity	
	Science	
	Technology	Technology drives social structures through models of feudalism, socialism, capitalism and post-modernity But diverse outcomes are possible from the same technology base, for example, capitalism in US and Europe but communism in Russia, and different social structures in Asian society Social culture affects technology – technocratic entrepreneurial cultures enable technology, as does a strong venture capital sector, strong IPR law, innovation systems, supportive government bodies, partnerships with universities, open innovation concepts

Continued

Table 4.2 Continued

Entity	Dependency	Nature of dependency
	Productivity	Technology drives social structures mainly through the surpluses created by productivity, releasing labour to new roles and structures, and through the new configurations of labour needed for high productivity production
	Economy	
	Society	
Humanity	Nature	In its naked state, humanity is entirely dependent on nature. Nature's impact on humanity can be benign or hostile, but this is random, and the outcome is often harsh for humanity
	Humanity	
	Science	
	Technology	Humanity is hugely dependent on technology. Survival at all in a hostile natural climate is impossible without the mediation of technology. Survival of several billion human beings at above subsistence standards of living requires very extensive deployment of technology. Humanity is so intrinsically dependent on technology that humanity is not simply humanity but quintessentially 'techno-humanity'
	Productivity	Since time is a fundamental economic resource, it is through productivity that humanity's dependency on technology is effected
	Economy	The market predominates as the current determining mechanism of the economy. Market is an artefact and can have independent impact on humanity.
	Society	Society is an artefact. Its totality is greater than, or at least different to, the sum of its parts, and so it independently affects humanity which is its sole content and creator

The point here is that each bipartisan impact and dependency between any two entities does not only operate bilaterally between these two entities, but has to operate through a network exchange. This is because of the 'knock on' effect whereby a change in any one entity will affect all other entities directly, and then all other entities indirectly through the initial direct effects. So a change in technology will affect productivity, which will affect the economy, which will affect society, which will affect nature, which will affect humanity etc. But then those initial direct effects will cause second round effects back to each entity, including to the entity generating the initial change. Such complex multi-layered impacts will work through the system until a new stable equilibrium is found, at which the states of all entities are defined to harmonise with the states of all other entities. It is like a kaleidoscope.

What we do observe over time is that relatively stable configurations of the network become established to create an era in human society. We have explored above the historically observed configurations of the network, and the associated state of each entity which characterises each configuration of feudalism, socialism, communism, capitalism, modernity and post-modernity. There are some ways in which the entities are required to be in specific states for the existence of a stable configuration. For example, modernity requires high productivity generated from technology. But variations are possible so that the configurations are not entirely uniquely determined. Capitalism does have variants of its American, European and Asian types which exhibit differences in configuration, for example, in the associated social structure.

Three core questions remain:

- **What is the balance of power between entities in the network?**
- **Specifically does technology determine humanity?**
- **And so what?**

4.4 The balance of power

The entities within the model have been drawn to appear rather like planets in spatial configuration. Perhaps they mutually attract each other as in Newton's scheme of the planetary cosmos? Or perhaps they compete for power to determine the network? In the above interactions, which are indicative and illustrative rather than exhaustive, which interactions are the more powerful, the more regular, the more determining?

In considering which entities exercise relatively more power in determining system network outcomes, we can ask of each entity to what extent it is generally contingent, how different could each entity be, or has it been historically? On this basis, nature has changed in the land-use patterns imposed on it by humanity and human social artefacts, in the depletion of its mineral resources, the absorption of emissions, and the reduction in bio-diversity. These changes may be significant in causing further climate change, specifically warming of the earth's atmosphere with consequences for sea levels and sustainable life. Nature has however remained largely unchanged in the majority of its materials and processes. Science, technology and productivity have changed immensely in the last 200 years. They are contingent; they could have been very different and were very different. Social structures and economies have also changed noticeably. Humanity has changed from hunter-gatherer to post-modern consumer over a longer period of time. How far this represents a fundamental change in the intrinsic nature of humanity, or only in the context of humanity, is not clear.

Reversing the analysis, we can ask how far each entity has created impact on the network. Nature is overwhelmingly determining. It feeds all network outcomes. There is no technology which does not reconfigure a natural material or process. Humanity is weak in the face of nature alone. Whilst there are some areas in which human economies are virtual rather than real, the vast part of the economy is real and therefore impacted by nature. Nature therefore exercises huge impact, but is relatively little changed itself. It therefore has a high power-factor rating when network power is measured by the ratio of impact exerted over impact suffered.

In the following table, we estimate subjectively the relative impact each entity is judged to have exerted into the systems network. We then estimate, equally subjectively, how contingent each entity is relative to other entities in the network. We finally report a ratio of impact exerted over impact suffered as an indicator of the relative power of each entity in the network compared to the other network entities.

Table 4.3 Impact ratings in the model

Entity	Impact exerted = E	Impact suffered = S	Network power ratio = E/S
Nature	Very high	Moderate	Very high
Science	High	Low	Moderate
Technology	Very high	Low	High
Productivity	Very high	Low	High
Market/ economy	High	Low	High
Society	Low	High	Low
Humanity	Moderate	High	Low

For fun and potential further enlightenment, let's try attaching some weights to this guesswork. If we set Low=2, Moderate=3, High=4 and Very high=5 we get

Entity	Impact exerted = E	Impact suffered = S	Network power ratio = E/S
Nature	5	3	1.67
Science	4	2	2
Technology	5	2	2.5
Productivity	5	2	2.5
Market/economy	4	2	2
Society	2	4	0.5
Humanity	3	4	0.75

Sorting this by network power ratio, we get

Entity	Impact exerted = E	Impact suffered = S	Network power ratio = E/S
Technology	5	2	2.5
Productivity	5	2	2.5
Science	4	2	2
Market/economy	4	2	2
Nature	5	3	1.67
Humanity	3	4	0.75
Society	2	4	0.5

On this inexact putative methodology, we get technology and productivity as the most determining power agents in the systems network, followed by science and the market economy. Nature itself is a weaker agent in the network, due to its passive role and its lack of a purposive function. Humanity and society trail with low determining power.

Although it is easy to scorn this methodology, it is in fact the inescapable way we implicitly think and philosophise about the real and virtual artefact world in which we live. As presented, the methodology is unrealistically, even ridiculously, exact, but we still do have to think around the relative weights we would ascribe to each entity and each interaction. What is either surprising and/or disturbing about the tentative analysis, is that humanity is ascribed a low power-rating. According to an initial attempt to evaluate the complex model of interactions, humanity appears less in charge than its cognitive active powers lead it to think. We are more subject to our real and virtual artefact context than it is to us. We are not, however, totally powerless. We may be more takers of our context than makers of it. Fatalism may therefore have some role to play in human understandings and religions. It does at least force some humility on us. Nevertheless, a positive quadrant does exist where we are makers of our context, proactive rather than passive. The truth, as ever, is that humanity is both a taker and a maker of its context, and therefore of its life. Attempting to understand the nature of the taking and making interactions with the other entities in the model is heuristic and enlightening, and a more refined understanding will enable us to adopt a more effective philosophy of humanity and of technology.

4.5 The original question

Specifically, we now have a more refined answer to the original question of whether technology determines humanity. We have seen that this question cannot be adequately addressed simply bilaterally as an issue only between humanity and technology, but has to be considered in the multilateral systems network model involving at least seven entities, of which two are real and five are artefact. This complexity may make the answer to the question of whether technology determines humanity even more difficult to answer. However, it does at least inform the answer.

First we ask generically whether and to what extent artefacts can determine humanity. Taking two other artefacts in the model, do market and society determine humanity? The answer is ambivalent. Theoretically, the artefacts of market and society do not necessarily determine humanity. But in practical reality, they can and do. Theoretically, we could establish an overt agency to decide every economic and social issue for us. But in reality, this has already been tried in models of the planned economy and state communist society, and has relatively failed. So we allow market and society free scope of action, and we agree to unleash processes that may and do constrain us, since micro communal decision-making is too inefficient. We implement a trade-off. Can we then wheel back from this position at any time when its outcomes are undesirable? Probably we can, but the effort involved to adjust outcomes may still be judged excessive. Sheer systems inertia all too often wins and is determinative. Like Kafka's characters, we are spun around by the social and market artefacts we have created, and feel powerless to change them or their outcomes. Political lobbying is in vain, action politics is difficult to organise effectively. Except that, once again, technology is rising to the cause as we have seen social networking technology manage effective social change movements in recent Middle East politics.

In order to synthesise an answer to the specific question of whether technology determines humanity, we can take various strands of data from our earlier description of the network in operation, and evaluate each noted phenomenon for its interpreted meaning for technology's determination of humanity.

How can we summarise this list of observations into an analytical statement and understanding? Our data is unstructured, random and biased. Nevertheless, from the phenomena which are observed in the data, certain key interpretations become common. Firstly, that technology is objective. Humans can develop technology but they can only develop technology which is objectively possible and available. Secondly, that we need to distinguish **technology in existence** from **technology in application** when we make an analytical or philosophical statement about technology. For example, technology may be totally objective existentially, but market contingent in application.

Table 4.4 Phenomena and technology determinism

Phenomenon observed	Implication for Technology Determination
Human beings could not survive without technology	Technology determines humanity
Contraceptive technology prevents human reproduction	
Medical technology extends human life expectancy	
Medical technology improves human health outcomes	
A hydrogen bomb could extinguish human life	
Technology can have unintended outcomes	Humanity cannot wholly control technology
Impossible to foresee all likely future outcomes of a technology to inform a decision on it	Humanity cannot calculate technology in order to manage it
Military technology determines political and social outcomes	Technology determines society
Technology is only and always developed by humanity	Humanity determines technology
All technologies have diverse potential applications	
Science was right about roof bolting in mines	Correct science (correct knowledge about nature) is objective and not subject to social construction
Powerplant emissions are reduced by a combination of technology and political will	Applied technology outcomes are politically contingent. The technology itself is only partially politically contingent since it has to be possible, that is, technology is partially objective
FGD and SCR are not socially constructed	Technology is in part objective and not socially constructed
Thixotropic aluminium was market contingent	Objective technology is market contingent in application

Continued

Table 4.4 Continued

Phenomenon observed	Implication for Technology Determination
GM food technology is politically contingent	Technology is partially politically contingent
Transport technology changed the space-time existence of humanity	Technology determines humanity
Hitler invaded Norway to get magnesium for the VW Beetle car	Politics is technology contingent
The competitive market determined the applications for aluminium vs magnesium, and aluminium vs copper	Objective technology is market contingent
US car emission legislation led to gains in market share for Honda's CVCC engine	Objective technology is politically and market contingent
Price/performance criterion determines technology market success	Objective technology is market contingent
Land use geography determines high speed train technology deployed	Objective technology is geo-economically contingent
Consumer choice of modal transport technology is time related	Objective technology is market contingent
Concorde airplane supersonic transport technology was discontinued	Objective technology is politically and market contingent
Deployment of malaria control technology is constrained by aid agency budgets	Objective technology is politically contingent
Eucomed data shows large impact of medical technology on infant mortality and life expectancy	Technology determines humanity
Technology cannot itself organise the convergence between technologies needed for any application	Technology requires human agency and is not autonomous
Computer technology displaced many clerical workers	Technology determines humanity
Telephone technology was held back by slow regulation	Technology is politically contingent
Large companies sometimes keep technology on the shelf	Technology is commercially contingent
E-mail technology was withheld for some time when it clearly bettered facsimile technology	Technology is contingent

Continued

Table 4.4 Continued

Phenomenon observed	Implication for Technology Determination
Technology as innovation is enabled by social innovation systems between industry, government, universities, venture capital, IPR law and by 'technocratic' entrepreneurial culture	Technology is socially contingent
Boss/secretarial culture determined office technology deployment and vice versa	Technology is socially contingent
Powerplant emission legislation forced the deployment of FGD and SCR technology	Objective technology is politically contingent
Best Available Technology Not Entailing Excessive Cost became the protocol for environmental technology deployment	Technology is objective, but politically and commercially contingent
Extensive privatisation of infrastructure technologies such as power generation, rail etc	Technology migrated from political to market contingency
Smartcard technology faced market criteria	Objective technology is market contingent
Military technology is determined by government and feeds commercial technology	Technology is politically contingent
Technology can be marketed	Technology is market contingent
Technology creates food surpluses which generate economic and social change	Technology determines society
Technology defines workforce and population organisation	Technology determines society
Military technology is deployed differentially to social groups and nation states	Technology determines political power
Technology leads to consumer alienation from production	Technology determines humanity

Beyond these leading points we see that common interpretations from the phenomena are

- technology determines humanity
- technology needs human agency, humanity determines technology
- technology both in existence and in application is often market contingent
- technology determines society
- technology determines political power
- technology is politically contingent

The answer to our original question about whether technology is autonomous and whether it determines humanity, is therefore more complex than a simple affirmative or negative answer. Truth is usually nuanced.

Our analysis challenges traditional enquiry into the philosophy of technology in two ways

- The question of technology determinism cannot be answered uni-directionally
  - Technology does not act on humanity, or on any other artefact such as society uni-directionally, but the relationship is always bi-directional
- The question of technology determinism cannot be answered bilaterally
  - Technology does not act on humanity in an isolated bilateral relationship, but only through a more complex multivariable network of several artefacts

4.6 The main direction

Whilst the model is therefore multilateral and always bi-directional, we see that some power relationships predominate. The diagram on the facing page reconfigures the model to align to the main lines of determination

In this dominant flow version of the model, the materials and processes of nature are reconfigured by the intentionality and agency of humanity to effect technology existent in potential. Sometimes this is done via science, by 'knowing how', but sometimes directly by 'knowing that'.

Existent technology is therefore contingent on nature and humanity.

In order for the technology to reach application status, it has to either

i) be approved by society through its governmental institutions to be applied as military technology or

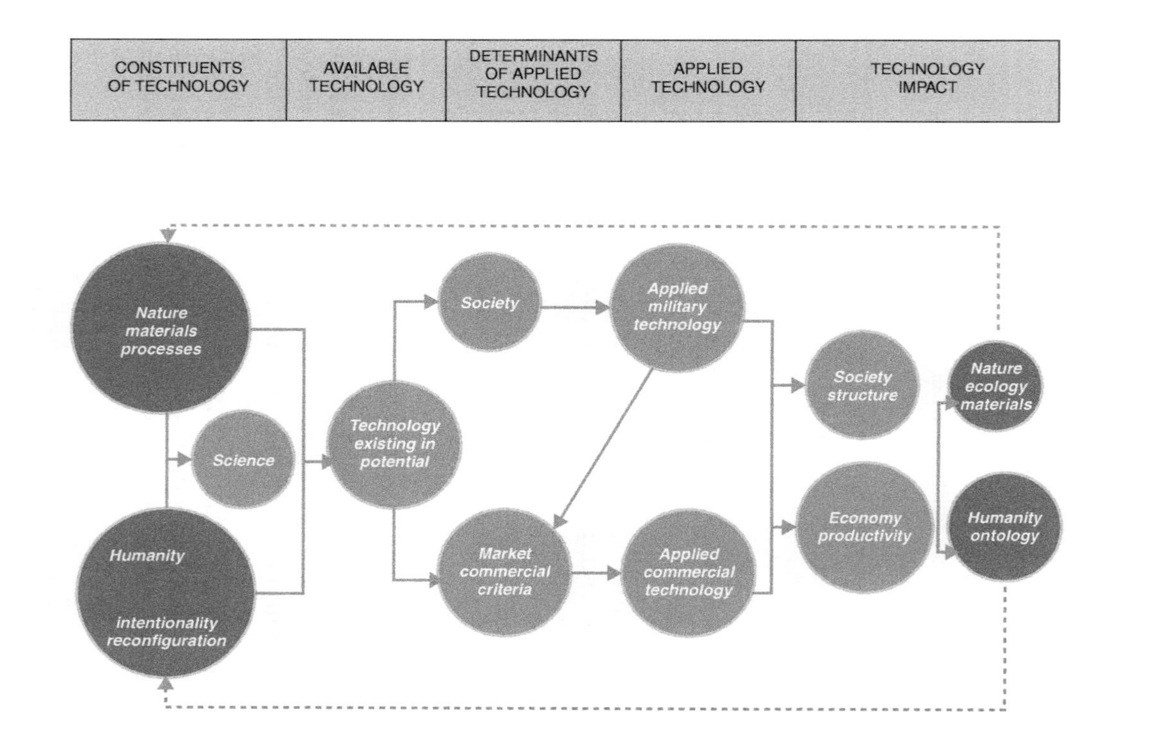

Figure 4.9 The main causal direction

ii) pass the market criteria of a positive downstream business case, a positive competitive price/performance positioning against other technologies, and a viable value chain to market

Frequently, applied military technology can be referred to the market criteria to test for commercial application.

Applied technology is therefore contingent on society and market artefacts.

The portfolio of military and commercial applied technologies then determines the economy via the lever of productivity and social structures, in the way described in Section 3.2.4.

Society and economy are therefore technology contingent.
Nature's ecology and humanity's ontology are impacted and are therefore also technology contingent.

4.7 So what? Implications of the model

There is an established methodology in business presentations. The presenter works through Powerpoint slides with sparsely worded bullet point statements, accompanied by a few core graphs and diagrams, on the basis that one picture is worth a thousand words. Academics snort in derision at these formats but could well learn from them, as the alternative is lengthy text, liberally peppered with numerical data, nested within sentences. Soviet Union reports were always presented in this latter way and were all the less comprehensible for it. Philosophy in particular would become more widely communicable if it adopted the despised bullet point.

The contemporary business presentation has to be short to retain audience attention, but crucially has to address the 'So what?' question at the end. Analyses must have implications. Strategies must be formulated, actions proposed, business plans and risk assessments calculated.

We therefore now consider the 'So what?' challenge to the exposition and development of our systems network model and philosophy of technology.

There are three dimensions of reaction to the exegesis of the model:

- awareness
- disposition
- action

Awareness is a mixed blessing. Consciousness defines humanity as unique and conveys advantage, whether in terms of inclusive fitness for survival of the species, or as an exaptation adding intellectual and emotional sensitivity to engage literature, music and art. However, as Sebastian Faulks comments in his novel 'Engleby', self-awareness is also a curse, forcing uncertainty, generating issues of identity, and the need to fulfil some sense of purpose onto humanity. There is a spectrum of responses. Some manage to live with low awareness settings, others struggle with neuroses generated from their awareness, whilst others power their awareness up, becoming untroubled achievers.

Similarly with a philosophy of technology. We can dull our awareness and simply live. There is a sense in which this has actually been the dominant reaction of humanity to technology and to a philosophy of technology. We seem content to drift along, or to ride the wave. We may feel very active in developing and applying this technology, but we are less interested in an awareness of how it is operating on us. Writing this book indicates my own urge for us to become more aware of the process of technology, more aware of its ontology, more aware of its symbiosis with humanity, more aware of whether and how it defines us. This book has sought to develop a nuanced answer to these questions.

If we choose to engage with this awareness, then what of our options for disposition and action? In the first place, awareness partially determines disposition. If we are unaware, then there will be no defined disposition or action. The content of our awareness more accurately determines our disposition. If we come to the conclusion that technology is autonomous, that humanity cannot resist its progress, then we may conclude further that any such resistance is futile. Our disposition may then be a shrug of the shoulders, and our action zero. If, however, awareness leads us to conclude that technology is entirely dependent on human choice and initiative, or that society, the economy and humanity at least partially determine technology, then we have a logical option to adopt a proactive disposition towards technology. Whether we do or not depends on personal motivation, time, resource and our perception of our historical and potential future effectiveness. Disposition determines action.

There is nothing to be said about zero or low levels of awareness or disposition, except to challenge them. What options come from positive awareness and proactive dispositions towards technology? Critical appreciation and respect may be the most helpful response. It's a response which is generated differentially at different role points in the social structure. In what follows, we discuss response options at different role points, ranging from person through consumer, worker, voter, business and educational institution, to government and society.

4.7.1 Person/people response

The response of a person or people to an awareness of a network-systems philosophy of technology is complex. On the one hand we may feel grateful that technology has delivered us life and lifestyle, and confident that it will continue to extend its impact favourably. On the other hand we may realise how dependent we are on technology, and therefore feel vulnerable. We may even feel threatened. Will technology's benign impact continue, grow, or decline? Will it continue to enrich us, or fail to deliver ever-increasing productivity and so impoverish us? Should we err on the side of caution and restrict our family size to less than the average 2.4 children? Or will technology's potential malignant characteristics predominate? Will nuclear, agricultural, or climate catastrophe overwhelm us, even destroy us?

These responses refer to our external context. But what about our identity? Descartes told us that our identity derived from our consciousness: cogito ergo sum – I think therefore I am. Now we see that, naked in nature without technology, not only is our survival unsure, but our identity is weak rather than affirmed, assured, confident, or triumphant. Our stronger, more confident, even proud identity is entirely due to technology. This should rather humble us. What rights we think we have, what achievements we have made, what confidence we have ... is all the gift of technology. We may respond that it is we who have developed this technology, but this pride and confidence, which is typical of modernity, forgets the contingent nature of technology. Technology is not simply determined by human effort, but also by nature ,whose materials and processes technology reconfigures. We can only discover technology, not invent it, although a creative intelligence is still needed for its discovery. We cannot do what technology cannot do, for example travel faster than the speed of light. And even if we, usually collectively rather than individually, may take some pride in our technological achievements, can we be so cocksure that the technology process will continue to grow exponentially, rather than asymptotically reach some constraint? The car that I drive, the clothes I wear, the culture I enjoy – all depend on technology, and the human contribution to this technology is usually someone else's rather than my own. I am as powerless to stop technology going wrong and delivering its malignant potential as I was to get it to go right and deliver the standard of living I enjoy.

Whilst cautioning against this over confident, even triumphalist, view of technology, its opposite is equally unsatisfactory. Within some university philosophy departments, a purely cynical denigration of technology can often be found. I attended one such seminar at a leading UK university: the lecturer, who is well known in his field, lifted his hands in despair and asked the student audience, of technology, 'Is it getting any better?' He and others discount the massive leaps forward in reducing infant and maternal mortality, in combating disease, and in raising standards of living from subsistence levels if they want. But it's noticeable that these cynics usually enjoy an above-average consumption profile, and often travel to several countries in any one year.

So a nuanced person/people response to technology might be – recognition of our dependency on technology, gratitude for this, humility in our identity, responsibility to engage with technology, to apply it to maximise personal productivity, and to act to minimise its negative impacts on the environment. We thus recognise reality. One way we can exercise this response is as a consumer.

4.7.2 Consumer response

We have established the market contingency of technology. In social democracies, the artefact of market determines technology paths. Technology has to generate a downstream business case in the market, demonstrate that it will generate new sales revenues in excess of its additional costs. It must also outperform the price/performance position of other technologies for the same functionality. And it must have a viable value-chain route to market available to it. Since technology is more market contingent than politically contingent, consumer response is more powerful than voter response. Consumption can be a blithe mindless act. But it can also be aware, informed, responsible and proactive. Cheap food, cheap energy, and cheap fuel are the deliverables of technology. Consumers campaign for low real prices, and so implicitly accept the technology that delivers them, usually without question.

Informed consumption however will be aware of the three elements of the price of the products they are buying: what element of the price signals the value of the product, what element codes for the technology by which it was made and delivered, and what element represents the income of the producer relative to that of the consumer. There has been a growing awareness of these issues in consumption. Whilst low food prices please western consumers, who often suffer obesity from over-consumption, the more recent concern is that the low price disguises an environmental cost externality of the technology that has been unpaid, and that the low price indicates unacceptably low producer incomes from labour-intensive production. The price derives from a technology which has failed to account for ecology or for producer income. The consumer has some option to decline this technology, by exercising a purchasing preference in favour of 'fair trade' products which guarantee higher producer incomes, and low carbon footprint products which have paid or reduced ecological costs. In order to exercise such a choice, effectively making technology dependent on humanity, consumers need accurate information and analysis. It is often assumed, for example, that loose fruit on sale in supermarkets has a lower carbon footprint than packaged fruit. But this is not always the case, since a complex set of factors determines the best environmental option. Similarly, it is essential to know that a higher 'fair trade' price does raise producer incomes, if that is the consumer's aim, and is not lost in higher per-unit overhead costs for the fair trade organisation. It is of course difficult if not impossible, to make such informed purchasing decisions about every product bought. Consumer 'watchdog' organisations therefore have an important role and greater potential in monitoring and reporting environmental and producer-income aspects of goods and services offered in the market economy.

Consumers can choose which product, produced and delivered by which technology, they prefer to purchase. This affects the technology's 'downstream business case' by affecting its sales revenues, and so determines its deployment. Similarly, consumers can choose whether to purchase on-line rather than in physical shops, and so affect the value chain, the route to market of the technology and its derived products.

If consumers choose to pay higher prices to represent environmental costs and higher producer incomes, then the consumer is deciding to take a lower real income and standard of living for themselves. This is arithmetically inevitable, but rarely realised explicitly. In the context described so far it is a micro decision. But consumers can act to determine a technology's market contingency in a more macro context. A simple decision to consume less in aggregate is theoretically possible, but rarely practised. Equally, consumers as travellers can decide to consume less internal combustion engine technology by preferring public transport, walking, cycling, or not travelling, to car travel.

Since technology is market contingent, people as consumers have technology choices. They need to exercise them.

4.7.3 Worker response

It took Keynes to point out the obvious; that most workers are also consumers, so that to reduce wages in the economy would also reduce demand and lead to recession. In a skill led economy, consumers rarely consume their own output of goods and services. As a result, we are all interdependent on each other for the productivity of our working production, and the standard of living this offers, both to workers in higher real wages and to consumers in lower real prices. This higher real wage, lower real price nexus, an apparently virtuous circle, derives from technology via productivity, an effect which needs to be more consciously realised.

As workers, we engage technology directly. Information and communications technology eases clerical and managerial tasks, reducing the time taken to accomplish them, increasing their accuracy and their extent. Thousands of supplier offers can be searched in Internet based procurement systems, simultaneously reducing the administrative search cost and identifying best supplier price/performance offers. Automated production technology drastically reduces the number of workers required to achieve any given level of output, and usually increases their skill requirement. Productivity may initially seem to threaten employment, and from the Luddite movement onwards there has been a tendency to resist technology-led productivity growth. Productivity is however the unavoidable necessity of higher standards of living. Gains in productivity at the microeconomic level of the firm have to be matched with other microeconomic new business development initiatives, or with macroeconomic expansion of the production economy. The labour capacity released by increased productivity has to be engaged in fresh production. The work ethic, associated with the Protestant religion by Max Weber, is a fundamental driver of the capitalist economy, so that the adoption of technology to raise productivity is what business strategists rightly call a 'win-win' strategy. Such productivity does not simply require capital investment, since capital equipment has to be matched by 'investment in human capital'. Worker skills and working practices have to be exactly aligned to the automation technology to derive the productivity benefit. Productivity is a systems concept. Simple technology transfer of advanced automated equipment into cultures without high productivity work skills, targets and practices will not achieve the productivity potential.

The simple arithmetic, that higher productivity through the deployment of technology necessarily leads to increased business profitability and social prosperity, is widely understood conceptually. It is less embraced practically. It should lead all economic policy, all individual action, all industrial effort. Instead, firms seek the results of profit, worker-consumers seek high real wage outcomes, without always focussing on the productivity which drives both. The result of this focus on the derived phenomenon of profit and real wage, rather than the causal factor of productivity, leads to a classic struggle over distribution of the economic product, rather than its initial creation. There is more heat generated over the question of how the cake should be divided, than of how big the cake itself is and could be. As a result the technology driver is paid less attention, and the aggregate outcome for all is less.

The Russian and Indian economies currently provide examples of this. In mining and manufacturing industries in Russia, productivity is well below best world practice. The productivity of Russia's coal mining sector is on average less than 1,500 tonnes/man year, well below international levels, which are six times greater. Some Russian coal mines, with productivities of some 7,000 tonnes/man year, do approach international comparisons, demonstrating that increased productivity is possible. However, even allowing for differential geological conditions and other adverse factors, it is industrial management and worker practice which has led to this low productivity, low wage equilibrium. There are cases where advanced technology is implemented but not used because it would raise productivity and reduce employment. For example, longwall mining equipment is capable of advancing automatically as the shearer cutting the coal senses that the coal has been cut and the whole 200–300m-long set of face equipment should advance. This dramatically reduces the number of workers needed to operate and advance the equipment set, and for this reason there are cases of shearer-sensitive automatic advance longwall sets working with this facility disabled in order to retain higher employment. The unavoidable concomitant of low wages appears not to be fully understood in this reasoning. In many industrial sectors, Russia faces the choice between low productivity low wage, and high productivity high wage outcomes, and seems to have actively preferred or else defaulted to the former. Only the high oil price saves the day for incomes and standards of living.

In India, the retail sector is hopelessly unproductive. Huge numbers of small shopkeepers maintain the traditional technology of sitting alongside each other with the same undifferentiated commodities at the same market clearing price. This inevitably drives shopkeeper incomes to subsistence levels. It also means that the food chain suffers from lack of adequate storage and refrigeration, leading to the loss of some 30 per cent of food produced. Well-organised well-capitalised supermarket operations from developed economy models are the obvious solution to this, and cannot fail to deliver a higher standard of living. The political resistance to this is, however, huge. Shopkeepers have votes and politicians vow to maintain their present extremely low productivity, extremely low wage outcome. This conservative traditionalism is characteristic of patterns of Indian social change where the old does not morph into the new, but the old remains whilst the new is added, leading to concurrent existence of historical layers of social practice. Ghandi led India well, politically, to its deserved independence, but his insistence on the preservation of traditional artisan technologies, with local village spinning machines producing homespun for the clothing Ghandi wore, exemplified a retention of a low productivity technology which could only offer low standards of living to a rapidly growing population. He was right politically, but his understanding of technology and economics was flawed.

Business managers and shareholders need to fully embrace the core point that investment in technology for productivity is the generator of profit, and workers in all types of employment need to embrace the same point – technology and high productivity is the sole source of high real wages. This not only applies to the optimal use of equipment, but to working practices too. All aspects of every task offer a wide range of options as to exactly how they are undertaken. In each case, insisting on the highest productivity working practice will spread higher income throughout society. The private sector of the economy knows this more clearly, since failure to operate at competitive levels of productivity inevitably leads to decline and likely bankruptcy. In public sector operations, the same discipline does not bite, and so it is noticeably easier to adopt lower productivity working practices with higher numbers of people engaged to achieve functions. It is this imperative which has led some politicians to contract out former public sector operations to the private sector. Critics argue that this always leads to lower wages for those employed, but whilst this can be the case, it is not necessarily so if private sector operators attune their operations to be more efficient in terms of person hours per function, due to their experience of the more competitive environment.

Marx called for the workers of the world to unite, since they had nothing to lose but their chains. His call was justified given the conditions of working class employment documented by his colleague Friedrich Engels in his 'The Condition of the Working Class in England in 1844'. Capitalism had not then generated sufficient productivity, and politics had not promoted a more equal income distribution. A more relevant motto after the failure of state-managed planned economies, where workers and citizens were enchained, is for workers to unite in embracing and committing to maximum productivity in technology and working practices. If this is supported by politically determined acceptable income distribution, then all will gain from commitment to high productivity.

4.7.4 Voter response

Some technology is determined in the public domain, is politically contingent. This is the case for

- military technology
- technology for public sector application
- government initiatives in fostering 'national innovation systems'
- specific science research funding
- very large technology projects
- legislation enabling or constraining technology

In these areas, voters have some influence on politicians. This influence is rarely directly on any one of these issues specifically, since most democracies operate representative democracy models rather than issue specific referenda. However, an open engaged critical press and media is able to address these issues on a regular basis with the political class, and voters can then factor these issues into their overall voting preference at elections. In Section 4.7.2 we considered the response to technology of consumers, and argued that, with increasingly privatised major sectors of the economy, the response as consumers is more regular, more wide reaching, and more effective than their response as voter. Democratisation of technology can be as much via the market as by the political election or some political process. Technology is so atomised and continual in the way it reaches us, that the market is a more effective mechanism for micro-decisions than a political process, which would struggle to cope with the myriad combination of technologies being offered, matched by the myriad political points of view. There would be a very long queue of technologies awaiting consideration politically – sclerosis would result.

Military technology is more consolidated and identifiable. Its end technologies of weaponry, intelligent systems, and transport are designed to deliver fatal attacks on infrastructure and on people. Its portfolio is extensive, including nuclear bombs, land mines, armour piercing bullets, satellite guided cruise missiles for claimed precision attack, supersonic aircraft, submarine launched missiles, and the ever more sophisticated and apparently ubiquitous gun. Even mediaeval military technologies, bows and arrows, and later the cannon ball, were devastating in their effect. The constituent technologies of military technology are wide ranging, from advanced materials technology such as breathable fabric for clothing, lithium alloys for structures, to advanced computer algorithms and communications technologies. It is often claimed that the 'spin off' from these constituent military technologies represents a huge contribution to the commercial civil economy, and that it is only the determined nature of military technology development which could have added these constituent technologies. The potential for military technology is even wider. Chemical and biological weaponry could prove more effective and deadly in combat than even nuclear bombing.

Military technology is and should always therefore be politically contingent. More debate about the choices society faces in developing, deploying and disseminating military technology is justified. International conventions such as the Geneva Convention do define which military technologies are allowed and which are proscribed. The moral foundation for such choices needs extensive separate treatment, since it is questionable whether human death is additive, whether one death matters less if it is one of thousands killed, and whether the mode of death is significant, or whether it is acceptable to be killed by a bullet rather than a bacterium. I should state my own conviction here that a single human death does matter just as much as if it were one of thousands, and killing by any means is just as reprehensible as by any other, but these are subjective issues requiring deeper debate. It is an indication and indictment of the inadequacy of the political process to control technology generally, that these issues of military technology find no regular forum for debate.

Technologies for public sector application classically include transportation infrastructure, healthcare, education, power generation, gas pipelines, telecommunications networks, water supply, sewage treatment and waste disposal. Even though some of these sectors have transferred to private sector supply in some countries, they are still the main macroeconomic sectors where technology choices arise.

Transportation technology choices are both inter-modal and intra-modal. Modal options for transporting people and products include road, rail, air and sometimes sea. Each of these has its own set of technologies for infrastructure, vehicle and operating system. Each of those technologies in turn has constituent technologies of car, lorry, train, boat, and within those, technologies of diesel, petrol or electric traction. Here there is frequent and emotive public debate. The irresoluble trade-offs are between the ecologically superior train and the consumer comfortable car, between the carbon footprint of air travel and the consumer commitment to take foreign holidays or wander the world on student gap years.

The economics argument is that car travel does not meet its infrastructure cost or its environmental emission cost, and that its energy cost understates the discounted price of imminent fossil fuel shortages. There are at least four and maybe more potential responses to this challenge to car and road technology. One is to impose infrastructure and emission costs onto car travel, and return the technology choice to the consumer, who would then be facing full comparative cost information. The second is to incentivise use of lower emission modes of transport, specifically rail. The third is to re-engineer car technology to overcome these challenges. And the fourth is to develop technologies such as video conferencing to reduce the perceived need to travel for personal contact.

All of these are in fact currently engaged as a response to the disadvantage of car transport technology. Car travel faces petrol taxes and road use tax schemes. Rail travel is frequently subsidised. New automotive technologies increase power/weight ratios of the car by optimal re-engineering using materials with best strength/weight/cost ratios such as aluminium, magnesium and composites. Traction technology urgently develops hybrid electric motors, and the potential for fuel cell power. Speed limits applied initially for safety reasons, also have an effect on fuel consumption. Virtual technology alternatives to travel are deployed to reduce physical travel.

Healthcare technology drives forward to offer treatment, cure and correction to ever-wider health issues. Medical technology can combat many bacteria but few viruses. It is increasingly successful against cancer. It routinely corrects heart malfunction, fits new hips and knees, and dispenses a very wide range of pharmaceutical therapy. Of all technologies, medical technology is probably the one which has impacted human life the most. It has extended life expectancy, has increased the health experience of life, and has moderated or eliminated pain. The two political issues which arise are what proportion of economic resource and activity to allocate to healthcare, and how and to whom these should be allocated. The technology is capable of generating almost infinite demand, particularly in healthcare systems which are free at the point of use. Application of the technology is then necessarily rationed. Some medical technologies raise significant ethical issues, for example, in fertility technology, stem cell research, animal experimentation, and end of life treatment. Medical technology is less market contingent than other technologies, due to a prevalent ethical position that price, income and profit should not determine health outcomes, and therefore should not determine medical technology. Medical technology choices, both which medical technologies to develop and which medical technologies to deploy, are often political, but not democratic. They are made by specialist agencies, small cabals of government ministers and civil servants, or by doctors making pragmatic real-time decisions. Meanwhile, pharmaceuticals are developed almost entirely in the private sector, where their development is profit contingent. Overall, this seems to be a very obscure way to manage healthcare and medical technology.

Power generation technology is also fundamentally societal. Electricity has powered huge changes in human life, although a quarter of the world's human population still has no direct access to electricity. Steam turbine technology, with the steam generated either from coal boilers or nuclear fission, and gas turbine technology, often in conjunction with a steam turbine in CCGT configurations, are the two main sources of electric power, together with hydro turbines and wind turbines. Solar power and fuel cells remain marginal technologies. The main political considerations are fossil fuel depletion, atmospheric emissions, and safety. Combined cycle gas turbines operate at higher thermal efficiencies, and at half the equivalent emissions to coal fired steam turbines. Political action has secured reduction in SOx and NOx emissions to the atmosphere, with a consequent increase in the production cost of electricity due to installation of flue-gas desulphurisation and selective catalytic reduction technology. This has been an implicit political choice, since the populace was not consulted on the explicit trade-off. Nuclear power offers the low-emission solution adopted by France, but lobby groups oppose nuclear power on safety grounds. Once again, difficult trade-offs arise. In the UK, annual electricity consumption is some 350 TWhr which is generated from coal, gas and nuclear with some renewable. The political wish list is for adequate cheap power, without emissions and without nuclear, a set of desirables which are mutually exclusive. If technology is to resolve this dilemma of expectations, it will be by the development of clean coal technology, through ultra-critical coal fired boilers and CO_2 capture and sequestration, but this will further increase the production cost of a unit of electricity as SOx and NOx reduction technologies have done. Renewables technology also produces very high-cost electricity and has limited scope. So a short-term technology strategy, of – renewable gas CCGT plant and nuclear technology for power generation, is evolving, with the longer-term hope of a strategy with clean coal and nuclear for power generation, and gas diverted to premium applications such as domestic central heating and cooking.

The point of this discussion is that the technology choice is political. Power generation technology is politically contingent. However, the complex factors in determining a power technology strategy render the decision process too difficult. One almost suspects that neither populace nor politician wish to strike the trade-off, preferring to pretend that the full wish list of any amount of low cost zero emission electricity is fully available. It is not. Without an explicit decision, the technology strategy for power generation, like the health technology outcome discussed above, becomes implicit, and slips between the cracks. Technology may be politically contingent, but it is not readily politically determined.

Sewage treatment and waste disposal are also public domain technologies. Political decisions determine operating constraints, but these depend on the availability of technology. The European Union determined that untreated sewage could not be piped out to sea but must be treated and then deposited at sea from long pipelines. Sewage treatment technology developed to enable low footprint vacuum sealed treatment plants using advanced membrane and reverse osmosis technology to be located underground to preserve the local visual amenity. Technology and political decisions have to be coordinated.

Government management of science and technology includes the fostering of national innovation systems. Many governments make regular policy initiatives in their support for science and technology. This ranges from supporting pure and 'blue sky' research in university science departments, to grant aiding industrial research, or specific technology development. The policy process of 'picking winners', for example of a sector like biotechnology, or a specific technology like automated vision or fuel cell research can fail, since governments are not gifted at seeing commercial futures. Policy is therefore more commonly directed to fostering supportive and positive climates, giving incentives for venture capital activity, and drawing necessary networks together. Very large technology projects such as the CERN Large Hadron Collider, the human genome project, and space research and flight, require government support and so are politically determined.

Despite our emphasis on the consumer market contingency of technology, we see that wide ranges of technology are politically contingent. The lack of any democratic reference within the political technology determination process, means that in reality technology, which is politically contingent, becomes bureaucratically contingent. The political process can at best only refer a few mega technologies to the electorate, for example nuclear power, but even this is then part of a much wider election manifesto. Representative democracy gives political determination to elected politicians, but their inability or unwillingness to make technology choices means that apparently politically contingent technologies seep through the government process which proves rather porous to them.

4.7.5 Business response

Commercial business should have greater regard for Karl Marx and his insight that competitive business advantage depends on a firm's comparative technology portfolio. This can be along any of the dimensions set out by Joseph Schumpeter, new enhanced products, new more effective production processes, new or differentiated sources of supply, new markets, and new models of business organisation. Technology advantage in Schumpeter's sense is not therefore necessarily a question of science, although it does include science. Technology does enable new models of business organisation: Internet technology has enabled on-line sales, and viral marketing. The business which engages available technology and stimulates development of new technology in these applications, will succeed competitively against those who do not. Global markets are more transparent and accessible through Internet technology, allowing best supplier offers to be found, as well as trading platforms to be accessed. The firm which moves ahead will gain advantage.

Investment in advanced manufacturing technology requires long term vision and commitment. Funding needs to be long term. It is noticeable that companies from cultures with long term perspectives are able to invest to become and remain technology leaders in their sector. Siemens has the advantage of the long term view German investor market, and so can develop and deploy technologies with long pay back periods of ten years or more. In comparison, the short-term demands of London Stock Exchange investors, where payback periods as short as 18 months were expected, will rule out technology investment with necessarily longer payback periods. The result is the survival, success, and market domination of continental European firms in major sector technologies, Siemens, Philips, ABB, Alstom, Ansaldo, Alcatel-Lucent, Sony-Ericsson are examples, and the demise of UK companies such as GEC and Marconi, and all indigenous British car manufacturers. Arnold Weinstock, the chief executive of GEC famously ran the company by requiring set profit margins and return on capital employed from each of its divisions, and boasted that he never visited the company's factories. Rolls Royce's aerospace business bucked this trend and remains a world-leading technology company, together with other technocratic companies such as Renishaw, which was spun off from Rolls Royce.

Despite its core importance for business success, very little technology business management process is implemented in industrial companies. A technology alert company should

- operate management processes to constantly search global technologies

- prioritise potentially relevant technologies
- evaluate developing and available technologies
- source technologies from an optimal matrix, for example, licenses, alliances, universities
- run the technology market evaluation methodology of
  - downstream business case
  - competitive price/performance
  - viable value chain
- determine its own technology development and R&D budget
- proactively manage its resulting technology portfolio, applying, licensing etc
- continually evolve internal management organisation technologies
- optimally manage downstream technology market positions, market shares and margins
- value the company according to its technology portfolio

It is rare to see such an explicit technology business process implemented, even in the largest companies. Technology is too often confused with engineering and thought of as 'technical'. It is in fact the key determinant of company performance in a technology mediated world and market economy. Companies need to develop a 'technocratic' culture which requires a merger of technology, marketing, financial and organisation skills. This is a rare combination, certainly in any individual manager, but often also within a whole organisation or social culture. It is more often found in the European and Asian company model than in the more financially oriented Anglo-American model of the firm. In UK culture, deep understanding of the key advantages of a technology is rarely combined with an understanding of key market phenomena in the same person. Finance and marketing are one thing, technology, engineering and production another. The service market is clearly important, but competitive leading edge technology is essential to give a unique market advantage to the service offered.

The investor community is aware that technology is important to a company's performance and value, but nevertheless values companies on purely financial measures. Audited accounts show financial profit and loss and balance sheet data. Analysts offer a bewildering range of financial ratios from price/earnings to various cover percentages, 'acid tests' and 'quick ratios' as the key valuation of the company. All this is a historic approach to the valuation of a company, a look in the rearview mirror. In reality, a company's value is determined by a forward looking view, and this is more affected by its technology portfolio than by its financial measures, although these are not unimportant as they will resource the technology strategy. When companies are valued by financial measures alone, technology only enters the equation as one of several factors which influence the price/earnings ratio. It is not specifically measured and reported separately, and its core effect on company value is not sufficiently considered. Careful thought needs to be given to a methodology to extend company accounts to include technology portfolio valuations.

4.7.6 Education response

There is an informational, awareness and therefore educational gap in mediating technology to society. This gap is greater in some national cultures than in others. Cultures have status arbiters. In some cultures the arbiter of status might be wealth, prowess at sport, or aristocratic antecedent. In others it might be institutional power. In the surge of nineteenth-century modernity, it was often science and technology. Charles Darwin and Isambard Kingdom Brunel had heroic status in the UK, as did Marie Curie, Louis Pasteur and others in France, and other scientific researchers throughout Europe and the USA where modernity flourished. It is only in post-modernity that image has displaced content, and footballers receive greater acclaim than scientists or engineers. Technocracy has morphed into 'celebrocracy', just as democracy has eroded into bureaucracy.

The scientist, technologist, and engineer has become a commodity item in a culture where function is less appealing than status. Moreover, in these cultures, even science, technology, and engineering are highly differentiated in the status rating. Science is academic and intellectual, engineering is vocational, technology is lost somewhere in between. This distinction is most clear in UK culture, to its loss and detriment. The educational structure differentiates strongly between intellectual and vocational streams, which are organised in separate institutions. It is not therefore surprising that the two do not meet to create the necessary synthesis. Most other cultures, including European culture, American and Asian cultures, regard engineering more highly. Where there are educational institutions with a historic focus on technology in the UK, such as the former Manchester Institute of Science and Technology which merged into Manchester University in 2004, or Cranfield University, or Imperial College London, they often show more interest in adding financial management skills to their technology education base, than do other educational establishments in adding technology. The Said and Judge business schools are funded and founded at Oxford and Cambridge respectively, but there are no new technology schools of equal standing.

There is in fact little if any main technology focus in mainstream academic education in the UK. There are very few courses with technology in their title, and even those are the technology of something specific, rather than technology per se. The subject material of this book is not offered in taught courses in the education sector. As the opening chapter on the philosophy of technology showed, there is only a small corpus of literature on the subject, and few academic departments or practitioners.

This needs to change. For intellectual reasons, far greater coverage should be given to technology. It does significantly determine the human 'lifeworld', human ontology, and the human life experience. The systems network explored in this book is operating, and does have very great impact. We need to understand it better. Pragmatically, our life, lifestyle, and standard of living depend on technology. Technology should therefore be a major component of education. The history of technology, the nature of technology, the philosophy of technology, the practice of technology, specific technologies, technology business processes, the future of technology, the social impact of technology, the social contingency of technology, the economic effect of technology, production technology, environmental technology, organisational technology, science and technology, technology and engineering, should all figure strongly throughout the educational system. Currently they do not, as a result, our societies are moving blindly on auto-pilot, cruising through technology space.

4.7.7 Society response

The alternative is for society to develop a far more informed view of its symbiosis with technology. We have argued that technology determines society, at least partially. Our social structures, our social behaviour, our social possibilities, derive from technology. Urbanisation is a technological phenomenon. So is alienation. Technology has at once aggregated our societies into cities, and fragmented them into atomised individual units. It is now working further social change by rendering society virtual. And so is culture. Art and music are manifestations of technology, from the harpsichord to the electric guitar. Civilisation would not be very civilised without technology's enablement. The nobility of Rousseau's 'noble savage' is uncertain. We should surely prefer to be more aware of some entity, some artefact, which determined our societies to this extent?

Equally we have set out the behaviouralist view that society determines technology, through the relative effectiveness of its 'innovation systems'. Social cultural attitudes towards science, technology and engineering do affect the technology achievement and lifestyle outcome. The pattern of social institutions and their behaviour towards technology, – including businesses, all layers of government, the law, banks, investors, universities – should therefore be cognitively determined in a sophisticated society.

4.7.8 Government response

Governments should have a cogent technology philosophy, policy and programme. Government is an adopter of technology both in its military and civilian roles. It also sets the agenda in education, and fosters science and technology in academia and industry. It determines economic policy.

Public sector administrative functions should showcase best productivity technology. Many do, as is the case in the UK for example with the Driver and Vehicle Licensing Agency and the Identity and Passport Service. These agencies have implemented cutting-edge web systems to reduce labour and other operating costs. They demonstrate that public sector agencies can achieve high levels of service and efficiency, without necessarily having to resort to private sector outsourcing to benefit from the efficiency driver of competition. Indeed, these agencies can sometimes exceed the efficiency of private sector organisations, which can themselves become well established in their niche and so less competitive. Other administrative functions of government are lamentably slow, with low productivity and low service outcomes. This again partly derives from national culture. In some cultures, government agencies are seen more as authorities than as service agencies, in which case the urge to maximise service and efficiency is absent. In other cultures where the service ethic is endemic and infectious, government agencies also strive to excel, and so become early adopters of productivity and service enhancing technology. The downside of this extensive automation is the loss of personal contact, and the dominant pervasive interface with computerised equipment. Many people bank on-line and rarely visit their bank. On-line shopping reduces personal contact. Purchasing a train ticket from complex fare options on screen can be a technophobe's nightmare.

Government also needs technology strategies for military, transport, healthcare, power generation, and waste disposal. The issues concerning these technologies have been discussed above. Government needs to be more explicit about the issues and choices involved in these sectors. Which military technologies are considered ethical and effective has to be determined against which threats appear more imminent. Developing and sourcing military technology depends upon a view of the world in terms of alliances with common interests, or potentially hostile blocs. If military technology is to be sourced upstream from commercial suppliers, then the globalised role of those suppliers in other world markets has to be considered. Globalisation makes unique military advantage difficult to maintain. Technology export bans are only partially effective since they identify the area for competing nations to focus their research. The trade-offs described above in the choice of healthcare, transport, and power generation technologies should be set out and resolved explicitly. Currently they are not, and governments get elected on post-modern criteria of image, rather than criteria of policy content. The end result is that technology becomes contingent on bureaucracy – not politically contingent. This is a very obscure subjective process. Technology needs better attention.

Coordinating the innovation networks is an important government role. Determination is needed to ensure optimal interaction between science, industry, academia, banking, and the law. If venture capital is too risk averse, and operates with too short time horizons, then government can prime the financial system and undergird true venture capital funds, as the Bank of Japan did in the early stages of industrial technology development. At one stage in the UK, most so-called venture capital investment went into management buy outs of existing companies, rather than to the funding of innovative technology. Building a technology-aware entrepreneurial culture is a government option. Sponsoring annual technology fairs, such as Germany's Hannover Messe, or a contemporary equivalent of the famed 1851 Great Exhibition at Crystal Palace might be effective stimuli. Sponsorship of technology university departments, student places, and research is also an effective option. Science research grants are necessary technology resources. Governments could be far more determined in the totality of technology policy. Too often lip service is given to science and technology and it is relegated to a junior function. It requires priority policy attention. It is far more determining of the outcome of life for the population of any nation than is its finance ministry.

In its responsibility for economic policy, government should focus more on technology as the driver of the economy. This demands analytical focus on the real economy rather than the financial economy. Finance can only service and measure the real economy. An effective technology economics realises that productivity is the key to economic analysis. Productivity determines the macroeconomic demand/supply ratio. Essentially rising productivity will increase per capita supply and decrease employment, thereby decreasing demand in a wage economy. The effect of increased supply and decreased demand will be to lower inflationary pressure. At the limit zero productivity will yield infinite inflation, and infinite productivity will yield zero inflation. Most classic analyses of inflation however see it as determined by the money supply, or according to current central bank policy, by the interest rate. These views exemplify the apparent urge to analyse economic phenomena by financial causal parameters. But this is only a partial explanation of inflation : real economic phenomena such as productivity and resource capacity utilisation should also factor strongly into the analysis and therefore into government policy. Trying to manage an economy by acting on its financial indicators is like trying to drive a car by manipulating its dashboard instruments.

Real technology- and productivity-led economic analysis would also realise that current recession in world markets arises when productivity rises faster than real wages. This may lead to increased profitability, but if this is not directed into consumer spending, then demand will prove defective, and recession ensue. In an interview with the Financial Times in March 2011, the UK trade union leader Bob Crow made a very apt observation: 'if you have robots build cars, how are robots going to buy them?'. This expresses the key factor behind a technology-led economic crisis. If there is output without wages, then there is deficient effective demand in the economy in a Keynesian sense. Policy needs to recognise this factor and seek correctives, either in higher real wages, to match the increased supply arising from higher productivity, or in innovative measures such as a citizen's income. In an ultimate technology world, where huge output is made available with very low levels of employment, some such social income mechanism will be essential. The alternative is to allow economic recession to reduce GDP to a point where it matches real consumer incomes. Such policies which are typical of monetarist thinking, are an effective rejection of higher standards of living available from technology.

Similar thinking applies to pensions policy. Financial calculations of amounts saved into pension schemes and later drawn as pensions do not represent the economic reality. In that reality, current wage earners making pension scheme payments are in fact diverting resources to current pensioners. Policies to raise the retirement age will not solve the pensions funding problem if in so doing GDP is not increased. This is a real possibility as the marginal product of otherwise retired workers will be negative if they displace younger highly productive and highly effective workers and managers.

In the final analysis, it is in education that governments' role in technology is most crucial. A huge shift in the awareness of technology and the response to technology depends on its greater inclusion in the education curriculum. I hope that this book may have made a humble contribution to that aim.

5
Conclusion – Technology as Artefact and Artefacts' Effect on Humanity

A tentative conclusion is now drawn from the above analytical description of technology in a networked system. The various interactions described give substance and therefore justification to the opening assumptions of the model. The objective realities of nature and humanity do interact in a complex network through the artefacts of science, technology, productivity, the economy, the market, productivity, and political and social structures. Shifts can occur in any of these artefacts and any such shift will work its way through the network, redefining each artefact and crucially redefining the reality of humanity, and to some extent redefining nature itself. Science and technology are the artefacts which more often generate exogenous change into the network, and as such it can be claimed that technology does determine humanity.

The leading philosophical question from this relates to the ontology of humanity. Does technology indeed determine humanity, or generalising this, do artefacts determine humanity? The answer is at the same time both affirmative and negative. It is clear that artefacts can and do determine humanity. For example, the artefact of the market and particularly derivative financial market, is currently determining what humanity considers it can do in its economic life. Real potential human activity with real available resources is being constrained by considerations of debt in financial markets. This is not a necessary outcome. In the same way, the artefact of technology can and does determine human outcomes. For example, the current apparent military domination of the USA in its invasion of Iraq is purely the outcome of technology, as was the result of the war in the Pacific, as is the outcome of all wars. Might is indeed right, or rather might is the only right on offer. If the Taleban or North Korea developed superior weaponry technology, then the world would take on a different profile and adopt a different view of moral right.

Technology is close to determinative in cases where there is an efficiency gain with no other loss, that is, assuming *ceteris paribus*, a requirement which can prevail. In an industrial process, if a catalyst technology allows less energy and less raw material feedstock for the production of the same output and if this catalyst technology, for example the addition of an enzyme to the process, is of negligible cost and has no other deleterious effect, then it will almost certainly be adopted. It is in this sense autonomous. If one chemicals producer decides not to use the enzyme technology, then a competitor almost definitely will, enabling the competitor to market the same output at lower cost and price and win market share, forcing the initial producer to also adopt the technology. In this case the competitive market acts in conjunction with the technology to make adoption of the technology almost certain, almost determined. Humanity could rule against it, but lacks any reason to oppose and can also hardly notice and control the myriad of such small step changes which happen constantly. Further as set out above, technologies will work through commercialisation models checking their downstream business case, competitive price/performance, and viable route to market, and be implemented by osmosis into the market economy.

Technology is also determined in the sense of being constrained by the available reconfigurations of basic scientific processes existing in nature, although this might be a very large number of reconfigurations by permutation and therefore not a particularly binding constraint.

Nevertheless, the answer to the question as to whether artefacts in general and technology in particular can and do determine humanity is also negative. Humanity retains cognitive power and can, with sufficient determination, resist the power of the artefact it has created. Artefacts can determine humanity, but only if humanity is sufficiently unaware, docile, passive and supine to allow them to. It is possible to buck the market, and it is possible to decide against technology as Amish and other minority groups demonstrate. Whilst the Amish example demonstrates that human sub-groups can resist technology for themselves, it also demonstrates that they cannot resist it for others, that is, for humanity in total. The technologies of the industrial revolution – from steelmaking from iron ore, coal and alloys, to telephony, integrated circuitry, plastics etc – may have been resisted by the Amish, but found other compliant human groups as a channel to implementation. Similarly, GM food technologies and stem cell technologies may be effectively resisted by some human groups, but the likelihood is that other human groups will adopt them, thus rendering technology quasi-independent of humanity since a totally globally coordinated human response to technology is unlikely to be achieved.

If, however, humanity is to be able to moderate technology at all, it is essential to know how technology is at work. The aim of this book has been to clarify the modus operandi of technology through a systems network in symbiosis with humanity. A clearer analytical understanding of this systems network illuminates rather than alienates the technology process and, in so doing, sets humanity free in its relationship with technology. Our conclusion is therefore a clarified simplified Heideggerian proposition that

1. we should like to prepare a free relationship to (technology)
2. the relationship will be free if it opens our human existence to the essence of technology
3. when we can respond to this essence we shall be able to experience the technological within its own bounds.

Notes

1. Dusek, Val (2006) *Philosophy of Technology – An Introduction*, Blackwell p. 1.
2. Dusek, Val (2006) *Philosophy of Technology – An Introduction*, Blackwell p. 4
3. Berg Olsen, Jan-Kyrre and Selinger, Evan (2007) *Philosophy of Technology – 5 Questions*, Automatic Press p. iv.
4. Scharff Robert and Dusek Val (2003) *Philosophy of Technology, The Technical Condition*, Blackwell p. 222.
5. Berg Olsen, Jan-Kyrre and Selinger, Evan (2007) *Philosophy of Technology – 5 Questions*, Automatic Press p. 14.
6. Berg Olsen, Jan-Kyrre and Selinger, Evan (2007) *Philosophy of Technology – 5 Questions*, Automatic Press p. 23.
7. Berg Olsen, Jan-Kyrre and Selinger, Evan (2007) *Philosophy of Technology – 5 Questions*, Automatic Press p. 24.
8. Brey, Philip, *Feenberg on Modernity and Technology*, www.rohan.sdsu.edu/faculty/feenberg/brey.htm, p. 1.
9. Dusek, Val (2006) *Philosophy of Technology – An Introduction*, Blackwell p. 57.
10. Scharff Robert and Dusek Val (2003) *Philosophy of Technology, The Technical Condition*, Blackwell p. 211.
11. Berg Olsen, Jan-Kyrre and Selinger, Evan (2009) *New Waves in Philosophy of Technology*, Palgrave Macmillan p. 14.
12. Scharff Robert and Dusek Val (2003) *Philosophy of Technology, The Technical Condition*, Blackwell p. 212.
13. Berg Olsen, Jan-Kyrre and Selinger, Evan (2009) *New Waves in Philosophy of Technology*, Palgrave Macmillan p. 193.
14. Scharff Robert and Dusek Val (2003) *Philosophy of Technology, The Technical Condition*, Blackwell p. 213.
15. Scharff Robert and Dusek Val (2003) *Philosophy of Technology, The Technical Condition*, Blackwell p. 214.
16. Scharff Robert and Dusek Val, 2003, *Philosophy of Technology, The Technical Condition*, Blackwell p. 216.
17. Scharff Robert and Dusek Val (2003) *Philosophy of Technology, The Technical Condition*, Blackwell p. 218.
18. Achterhuis Hans et al. (2001) *American Philosophy of Technology, The Empirical Turn*, Indiana University Press p. 102.
19. Berg Olsen, Jan-Kyrre and Selinger, Evan (2009) *New Waves in Philosophy of Technology*, Palgrave Macmillan p. 207.
20. Berg Olsen, Jan-Kyrre and Selinger, Evan (2009) *New Waves in Philosophy of Technology*, Palgrave Macmillan p. 223.
21. Berg Olsen, Jan-Kyrre and Selinger, Evan (2009) *New Waves in Philosophy of Technology*, Palgrave Macmillan p. 245.
22. Berg Olsen, Jan-Kyrre and Selinger, Evan (2009) *New Waves in Philosophy of Technology*, Palgrave Macmillan pp. 247–248.

23. Scharff Robert and Dusek Val (2003) *Philosophy of Technology, The Technical Condition*, Blackwell p. 173.
24. Berg Olsen, Jan-Kyrre and Selinger, Evan (2009) *New Waves in Philosophy of Technology*, Palgrave Macmillan pp. 14–15.
25. Berg Olsen, Jan-Kyrre and Selinger, Evan (2009) *New Waves in Philosophy of Technology*, Palgrave Macmillan p. 84.
26. Achterhuis Hans et al. (2001) *American Philosophy of Technology, The Empirical Turn*, Indiana University Press p. 1.
27. Scharff Robert and Dusek Val (2003) *Philosophy of Technology, The Technical Condition*, Blackwell p. 398.
28. Scharff Robert and Dusek Val (2003) *Philosophy of Technology, The Technical Condition*, Blackwell pp. 398–404.
29. Scharff Robert and Dusek Val (2003) *Philosophy of Technology, The Technical Condition*, Blackwell pp. 386–397.
30. Berg Olsen, Jan-Kyrre and Selinger, Evan (2009) *New Waves in Philosophy of Technology*, Palgrave Macmillan p. 84.
31. Scharff Robert and Dusek Val (2003) *Philosophy of Technology, The Technical Condition*, Blackwell pp. 222–232.
32. Scharff Robert and Dusek Val (2003) *Philosophy of Technology, The Technical Condition*, Blackwell pp. 233–243.
33. Feenberg, Andrew 2010 *Between Reason and Experience*, MIT Press p. 5.
34. Scharff Robert and Dusek Val (2003) *Philosophy of Technology, The Technical Condition*, Blackwell pp. 265–276.
35. Scharff Robert and Dusek Val (2003) *Philosophy of Technology, The Technical Condition*, Blackwell p. 247.
36. Scharff Robert and Dusek Val (2003) *Philosophy of Technology, The Technical Condition*, Blackwell p. 327.
37. Scharff Robert and Dusek Val (2003) *Philosophy of Technology, The Technical Condition*, Blackwell pp. 252–264.
38. Smil, Vaclav (2005) *Creating the Twentieth Century: Technical Innovations of 1867–1914 and their Lasting Impact*, Oxford University Press.
39. Wikipedia, Martin Heidegger.
40. Scharff Robert and Dusek Val (2003) *Philosophy of Technology, The Technical Condition*, Blackwell pp. 405–412.
41. Scharff Robert and Dusek Val (2003) *Philosophy of Technology, The Technical Condition*, Blackwell p. 196.
42. Scharff Robert and Dusek Val (2003) *Philosophy of Technology, The Technical Condition*, Blackwell pp. 293–314.
43. Achterhuis Hans et al. (2001) *American Philosophy of Technology, The Empirical Turn*, Indiana University Press.
44. Ihde, Don (1990) *Technology and the Life World*, Indiana University Press.
45. Ihde, Don (1990) *Technology and the Life World*, Indiana University Press p. 17.
46. Ihde, Don (1990) *Technology and the Life World*, Indiana University Press p. 57.
47. Ihde, Don (1990) *Technology and the Life World*, Indiana University Press p. 26.
48. Achterhuis Hans et al. (2001) *American Philosophy of Technology, The Empirical Turn*, Indiana University Press p. 119.

49. Achterhuis Hans et al. (2001) *American Philosophy of Technology, The Empirical Turn*, Indiana University Press p. 123.
50. Achterhuis Hans et al. (2001) *American Philosophy of Technology, The Empirical Turn*, Indiana University Press p. 129.
51. Achterhuis Hans et al. (2001) *American Philosophy of Technology, The Empirical Turn*, Indiana University Press p. 133.
52. Achterhuis Hans et al. (2001) *American Philosophy of Technology, The Empirical Turn*, Indiana University Press p. 135.
53. Achterhuis Hans et al. (2001) *American Philosophy of Technology, The Empirical Turn*, Indiana University Press p. 138.
54. Achterhuis Hans et al. (2001) *American Philosophy of Technology, The Empirical Turn*, Indiana University Press p. 144.
55. Scharff Robert and Dusek Val (2003) *Philosophy of Technology, The Technical Condition*, Blackwell p. 260ff.
56. Scharff Robert and Dusek Val (2003) *Philosophy of Technology, The Technical Condition*, Blackwell pp. 293–314.
57. Scharff Robert and Dusek Val (2003) *Philosophy of Technology, The Technical Condition*, Blackwell p. 530.
58. Feenberg, Andrew (2010) *Between Reason and Experience*, MIT Press p. 6.
59. Misa, Thomas (2004) *Leonardo to the Internet, Technology and Culture from the Renaissance to the Present*, Johns Hopkins University Press p. 14.
60. Ihde, Don (1990) *Technology and the Life World*, Indiana University Press p. 164.
61. Scharff Robert and Dusek Val (2003) *Philosophy of Technology, The Technical Condition*, Blackwell p. 399.
62. Davies, Paul (2007) *The Goldilocks Enigma*, Penguin.
63. Popper, Karl (1990) *A World of Propensities*, Thoemmes.
64. Bohm, David (1984) *Causality and Chance in Modern Physics*, Routledge.
65. Skidelsky, Robert (2007) *Keynes – The Return of the Master*, Penguin p. xvi, p. 75,p. 78.
66. Schapiro, Stewart)2007) *The Objectivity of Mathematics*, Synthese.
67. Cartwright, Nancy (1979) *Causal Laws and Effective Strategies*, Nous. Hitchcock, Christopher, *Probabilistic Explanation*, Stanford Encyclopaedia of Philosophy.
68. Popper, Karl (2002) *The Logic of Scientific Discovery*, Routledge.
69. Curd, Martin and Cover, JA (1998) *Philosophy of Science – The Central Issues*, W W Norton p. 257.
70. Kuhn, Thomas (1962) *The Structure of Scientific Revolutions*, University of Chicago Press.
71. Curd, Martin and Cover, JA (1998) *Philosophy of Science – The Central Issues*, W W Norton p. 27.
72. Planck, Max (1949) Scientific Autobiography and other papers, New York pp. 33–34.
73. Kuhn, Thomas (1962) *The Structure of Scientific Revolutions*, University of Chicago Press p. 136.
74. Kuhn, Thomas (1962) *The Structure of Scientific Revolutions*, University of Chicago Press p. 136.
75. Kuhn, Thomas (1962) *The Structure of Scientific Revolutions*, University of Chicago Press pp. 171–172.

76. Kuhn, Thomas (1962) *The Structure of Scientific Revolutions*, University of Chicago Press pp. 146–147.
77. Kuhn, Thomas (1962) *The Structure of Scientific Revolutions*, University of Chicago Press p. 174.
78. Kuhn, Thomas (1962) *The Structure of Scientific Revolutions*, University of Chicago Press p. 94.
79. Mastermann, Margaret, *The Paradigm Concept*.
80. Curd, Martin and Cover, JA (1998) *Philosophy of Science – The Central Issues*, W W Norton p. 119.
81. Lakatos, Imre et al. (1970) *Criticism and the Growth of Knowledge*, Cambridge University Press.
82. Shapere, Dudley (1971) *The Paradigm Concept*, Science.
83. Curd, Martin and Cover, JA (1998) *Philosophy of Science – The Central Issues*, W W Norton p. 125.
84. Curd, Martin and Cover, JA (1998) *Philosophy of Science – The Central Issues*, W W Norton p. 128.
85. Kuhn, Thomas (1962) *The Structure of Scientific Revolutions*, University of Chicago Press.
86. Curd, Martin and Cover, JA (1998) *Philosophy of Science – The Central Issues*, W W Norton p. 239.
87. Curd, Martin and Cover, JA (1998) *Philosophy of Science – The Central Issues*, W W Norton p. 501.
88. Curd,Martin and Cover, JA (1998) *Philosophy of Science – The Central Issues*, W W Norton p. 433.
89. Curd,Martin and Cover, JA (1998) *Philosophy of Science – The Central Issues*, W W Norton p. 433
90. Curd, Martin and Cover, JA (1998) *Philosophy of Science – The Central Issues*, W W Norton pp. 685–719.
91. Curd, Martin and Cover, JA (1998) *Philosophy of Science – The Central Issues*, W W Norton p. 746.
92. Curd, Martin and Cover ,JA (1998) *Philosophy of Science – The Central Issues*, W W Norton p. 826.
93. Curd, Martin and Cover, JA (1998) *Philosophy of Science – The Central Issues*, W W Norton p. 865.
94. Curd, Martin and Cover, JA (1998) *Philosophy of Science – The Central Issues*, W W Norton p. 94.
95. Berry, Michael (2002) *Singular Limits*, Physics Today.
96. Curd, Martin and Cover, JA (1998) *Philosophy of Science – The Central Issues*, W W Norton p. 1064.
97. Ladyman, James (2002) *Understanding Philosophy of Science*, Routledge p. 6.
98. Ladyman, James (2002) *Understanding Philosophy of Science*, Routledge p. 230.
99. Arthur, W Brian (2009) *The Nature of Technology*, Penguin Books p. 18.
100. Arthur, W Brian (2009) *The Nature of Technology*, Penguin Books p. 193.
101. Smil, Vaclav (2006) *Transforming the Twentieth Century, Technical Innovations and Their Consequences*, Oxford University Press. Smil, Vaclav (2005) *Creating the Twentieth Century: Technical Innovations of 1867–1914 and their Lasting Impact*, Oxford University Press.

102. Smil, Vaclav (2006) *Transforming the Twentieth Century, Technical Innovations and their Consequences*, Oxford University Press p. 13.
103. Smil, Vaclav (2006) *Transforming the Twentieth Century, Technical Innovations and their Consequences*, Oxford University Press p. 43.
104. Bruland, Kristine (1989) *British Technology and European Industrialization*, Cambridge University Press.
105. Bruland, Kristine (1989) *British Technology and European Industrialization*, Cambridge University Press p. 111.
106. Ihde, Don (1990) *Technology and the Life World*, Indiana University Press.
107. Smil, Vaclav (2006) *Transforming the Twentieth Century, Technical Innovations and their Consequences*, Oxford University Press p. 205.
108. Smil, Vaclav (2006) *Transforming the Twentieth Century, Technical Innovations and their Consequences*, Oxford University Press p. 208.
109. OECD Health Data 2007.
110. Smil, Vaclav (2006) *Transforming the Twentieth Century, Technical Innovations and their Consequences*, Oxford University Press.
111. Pfaffenberger, Bryan, (1992) *Social Anthropology of Technology*, Annual Review of Anthropology.
112. Hughes, Thomas (2004) *Human Built World, How to Think about Technology and Culture*, University of Chicago Press.
113. Dosi, Giovanni et al. (1988) *Technical Change and Economic Theory*, Pinter.
114. Fagerberg, Jan (2005) The *Oxford Handbook of Innovation*, Oxford University Press p. 29.
115. Lundvall, Bengt-Ake (2010) *National Systems of Innovation*, Anthem Press p. 173.
116. Chesbrough, Henry (2006) *Open Innovation – Researching a New Paradigm*, Oxford University Press.
117. Fagerberg, Jan (2005) *The Oxford Handbook of Innovation*, Oxford University Press p. 1.
118. Feenberg, Andrew (2010) *Between Reason and Experience*, MIT Press.

Bibliography

Achterhuis Hans et al (2001) *American Philosophy of Technology: The Empirical Turn*, Indiana University Press.

Anderson, J L (2009) *Industrializing the Corn Belt*, North Illinois University Press.

Arthur, W Brian (2009) *The Nature of Technology*, Penguin.

Berg Olsen, Jan-Kyrre and Selinger, Evan (2007) *Philosophy of Technology – 5 Questions*, Automatic Press.

Berg Olsen, Jan-Kyrre and Selinger, Evan (2009) *New Waves in Philosophy of Technology*, Palgrave Macmillan.

Berry, Michael (2002) Singular Limits, Physics Today.

Bohm, David (1984) *Causality and Chance in Modern Physics*, Routledge.

Bohm, David (1980) *Wholeness and the Implicate Order*, Routledge.

Borgmann, Alfred (1984) *Technology and the Character of Contemporary Life*, University of Chicago Press.

Brey, Philip, *Feenberg on Modernity and Technology*, www.rohan.sdsu.edu/faculty/feenberg/brey.htm

Bruland, Kristine (1989) *British Technology and European Industrialization*, Cambridge University Press.

Cartwright, Nancy (1979) *Causal Laws and Effective Strategies*, Nous.

Chesbrough, Henry (2006) *Open Innovation – Researching a New Paradigm*, Oxford University Press.

Curd, Martin and Cover, J A (1998) *Philosophy of Science – The Central Issues*, W W Norton.

Davies, Paul (2007) *The Goldilocks Enigma*, Penguin.

Dosi, Giovanni et al (1988) *Technical Change and Economic Theory*, Pinter.

Dusek, Val (2006) *Philosophy of Technology – An Introduction*, Blackwell.

Fagerberg, Jan (2005) *The Oxford Handbook of Innovation*, Oxford University Press.

Feenberg, Andrew (1995) *Alternative Modernity*, University of California Press.

Feenberg, Andrew (1999) *Questioning Technology*, Routledge.

Feenberg, Andrew (2002) *Transforming Technology*, Oxford University Press.

Feenberg, Andrew (2010) *Between Reason and Experience*, MIT Press.

Higgs, Eric et al (2000) *Technology and the Good Life?*, University of Chicago Press.

Hitchcock, Christopher, Probabilistic Explanation, Stanford Encyclopaedia of Philosophy.

Hughes, Thomas (2004) *Human Built World, How to Think about Technology and Culture*, University of Chicago Press.

Ihde, Don (1990) *Technology and the Life World*, Indiana University Press.

Kaplan, David (2004) *Readings in the Philosophy of Technology*, Rowman and Littlefield.

Kirkpatrick, Graeme (2011) *Technical Politics : Critical Theory and Technology Design*, Bloomsbury.

Kuhn, Thomas (1962) *The Structure of Scientific Revolutions*, University of Chicago Press.

Ladyman, James (2002) *Understanding Philosophy of Science*, Routledge.

Lakatos, Imre et al (1970) *Criticism and the Growth of Knowledge*, Cambridge, Cambridge University Press.

Laudan, Larry (1984) *Dissecting the Holistic Nature of Scientific Change*, University of California Press.

Lundvall, Bengt-Ake (2010) *National Systems of Innovation*, Anthem Press.

Mastermann, Margaret, The Paradigm Concept.

Misa, Thomas (2004) *Leonardo to the Internet, Technology and Culture from the Renaissance to the Present*, Johns Hopkins University Press.

Misa, Thomas et al (2003) *Modernity and Technology*, MIT Press.

Mitchum, Carl (1994) *Thinking through Technology, The Path Between Engineering and Philosophy*, University of Chicago Press.

Okasha, Samir (2002) *Philosophy of Science – A Very Short Introduction*, Oxford University Press.

Pfaffenberger, Bryan (1992) *Social Anthropology of Technology, Annual Review of Anthropology*.

Planck, Max (1949) Scientific Autobiography and other papers, New York.

Popper, Karl (1990) A World of Propensities, Thoemmes.

Popper, Karl (2002) *The Logic of Scientific Discovery*, Routledge.

Prahalad, C K and Krishnan, M S (2008) *The New Age of Innovation*, McGraw Hill.

Rosenberg, Nathan (1992) *Technology and the Wealth of Nations*, Stanford University Press.

Schapiro, Stewart (2007) *The Objectivity of Mathematics*, Synthese.

Scharff Robert and Dusek Val (2003) *Philosophy of Technology, The Technical Condition*, Blackwell.

Shapere, Dudley (1971) The Paradigm Concept, Science.

Skidelsky, Robert (2007) *Keynes – The Return of the Master*, Penguin.

Smil, Vaclav (2005) *Creating the Twentieth Century : Technical Innovations of 1867–1914 and their Lasting Impact*, Oxford: Oxford University Press.

Smil, Vaclav (2006) *Transforming the Twentieth Century, Technical Innovations and Their Consequences*, Oxford University Press.

Spiegel, Der (1966) Heidegger interview, www.ditext.com/heidegger/interview/html

Wikipedia, Martin Heidegger.

Index

academic literature, on philosophy of technology, 6–40
Achterhius, Hans, 34
acid rain, 109
activism, 38–40
Adler, Alfred, 66
aero engine, 97–8
agency, 17, 47–8
agricultural technology, 83, 90–3
agriculture, 6, 20, 30
AIDS virus, 3
air pollution, 109
alienation, 132, 138, 196
aluminum, 87, 94–5
Amdahl, Gene, 105
anaesthetics, 101
analgesics, 101
anthropocentric world view, 134
anthropological definition, of technology, 12–14
anthropology, 138–9
antibiotics, 3, 91, 100–1
antidepressants, 101
antivirals, 101
applied science, 133
applied technology, 179
Aristotle, 140
Arkwright, Richard, 88
art, 196–7
artefacts, 8, 16, 27, 31–2, 36, 38, 41–3, 48–51, 71, 173, 201–3
Arthur, W. Brian, 82–3
artificial intelligence (AI), 14, 44–5
atomic theory, 80
Atrazine, 90–1
automation, 125
automotive technology, 94–5, 189–90
autonomy, 25–7

Bacon, Francis, 7, 19, 134
barium, 84
barter exchange, 121
BATNEEC principle, 110
Bayesian criterion, 63
behavioural economics, 146–50
Berry, Michael, 78–9
Bijker, Wiebe, 7, 27
Bohm, David, 56, 141
Bohr, Niels, 56, 65, 141
Borgmann, Albert, 7, 30, 34, 38
Bosch, Carl, 83
Bostrom, Nick, 12
Brey, Philip, 9–10
British industrial revolution, 147–8
British Labour Party, 119
British Rail, 127–9
Bruland, Kristine, 89
Bulgakov, Mikhail, 1
Bunge, Mario, 7, 16
business response, to technology, 192–5

capital, 113, 115
capitalism, 20–2, 25, 134
carbon capture and sequestration (CCS), 87, 111
carbon dioxide, 17, 110–12
Carson, Rachel, 6, 90
Cartwright, Nancy, 78
casting technology, 87
catalytic converters, 95
cause and effect, 63–4
Chadwick, James, 83
chemicals, 90–1
chemistry, 70
Chesbrough, Henry, 149
citizenship, 130–1
Clarke, G., 117
coal, 85, 109–10
COBOL, 105, 106
Cockroft, John, 83
combined cycle gas turbine (CCGT), 87
command economies, 115–21, 128, 157
communications technology, 106–8

communism, 119–21, 137, 138
comparative economic value, 120
competition, 157–8
computer-aided design, computer-aided manufacture (CADCAM) systems, 88
computer technology, 104–8
Comte, Auguste, 29
consumerism, 119
consumer response, to technology, 182–4
consumer society, 3, 49, 124, 143, 155
contraception, 2, 102
Copenhagen interpretation, 56, 65, 141
Copernican revolution, 68, 70, 71, 72, 73, 133
copper, 95
creativity, 82
crime, 140–1
critical rationalism, 67, 68
Crompton, Samuel, 88
culture, 36, 138–9, 196–7
cyborgs, 1, 14–15, 18, 45

Dantzig, George, 117
Darwinian revolution, 70, 74
Davies, Paul, 52
DDT, 90
democracy, 3, 21, 22, 28, 38–40, 134, 138, 143
democratisation, 129
Denison, Edward, 115
Descartes, René, 134
determinism, 19–25, 48, 55, 174–7
Digital Equipment Corporation (DEC), 105
diseases, 2–3, 101
Dretske, Fred, 78
Dreyfus, Hubert, 34
Dummett, Michael, 58
Dumont, Louis, 141
Durkheim, Émile, 140–1
Dusek, Val, 6–7, 10, 29
dystopia, 29–34

ecological technologies, 108–12
ecology, 45, 144–5
economic growth, 114–15, 147
economic policy, 199–200
economics, 3, 57, 112–15
economic society, 131
economy, 48–9, 112, 115–30
  forms of, 115–23
  global, 125
  productivity in the real, 124–30
  science and, 163
education, 143, 195–6
efficiency, 149
Einstein, Albert, 56, 65, 66, 69
electricity, 85, 109, 110, 190–1
electric motor, 84, 93, 97
electromagnetic field, 44, 53, 55, 57, 80
Ellul, Jacques, 26–7, 38
emissions, 108–12
empiricism, 79–80
empiricist phenomenology, 34–7
endogenous metaphysics, 134
energy, 54
Engels, Frederick, 22, 32
engineering technologies, 88–9
Enhanced Oil Recovery, 111
Enlightenment, 10, 21, 55, 70–1, 133–6, 143
environmental technologies, 2
Etzler, J. A., 125
European Union Air Framework Directive, 110
exogenous change, 50

facsimile, 107
falsification, 65, 67, 70, 77
Feenberg, Andrew, 8–10, 25, 28, 29, 34, 35, 38–40
Fermi, Enrico, 84
fertiliser, 83, 91
fertility technology, 2, 102
feudalism, 3, 20–2, 132, 133, 134
Feynman, Richard, 80–1
fields, 54–5, 80
financial crisis, 124–5
Fleming, Alexander, 23
flue gas desulphurisation (FGD), 87, 109
flying shuttle, 88
focal practices, 38
food production, 90–3

force, 54–5, 80
Ford, Henry, 143
forming technologies, 87–8
FORTRAN, 105
fossil fuels, 85, 87, 109
Freeman, Christopher, 146
Freud, Sigmund, 66
Frisch, Otto, 84
fuel cell technology, 98–100

Garden of Eden, 1, 35
gas, 93
gas turbine, 84, 97–8
Gehlen, Arnold, 12–13
genetically modified (GM) crops, 39
geo-political societies, 130–1
global economy, 125, 158
global warming, 110–12
government response, to technology, 197–200
gravity, 44, 53, 54–5, 57
greenhouse gas, 110–12
green technologies, 111–12
Grove, William, 98
growth hormones, 91
Grubb, Thomas, 98

Haber, Fritz, 83
Haber-Bosch process, 83
Habermas,, 38–40
Hahn, Otto, 83–4
Hale, Benjamin, 16
Haraway, Donna, 14–15, 34–5
Hargreaves, James, 88
health, 2–3
healthcare technology, *see* medical technology
Hegel, G. W. F., 140
Heidegger, Martin, 6, 7, 12, 13, 29–32, 36, 37–8, 48, 51–2, 82, 144, 163
Heilbroner, Robert, 22–5, 51
Heisenberg, Werner, 56
Hempel, Carl, 77–8
High, Thomas, 88
high speed rail, 97, 127–9
Hill, Christopher, 133
holism, 141
homeostatic control systems, 104
Hughes, Thomas P., 144–5
human capital, 113
human consciousness, 134, 180
human enhancement, 12–14, 18, 44–5
humanity
  nature and, 61, 164
  ontology of, 201–3
  science and, 162
  technology and, 1–4, 49–52, 81–2, 172–7
human rights, 21
Hume, David, 75–6

IBM, 105–6, 117
Ihde, Don, 34, 37, 235–6
implicit knowledge, 61–2
induction, 75–8
industrial revolution, 147–8
industrial society, 132–3, 143
infant mortality, 102–3, 114
innovation, 23, 108, 146–50
insecticides, 90
institutionalisation, 144
instrumentalism, 15–18, 45, 47, 73
Intel, 115
intellectual property rights, 115
intentionality, 15–17, 47–8, 82
internal combustion engine, 84, 93, 94–5
Internet, 23, 38, 107–8, 137

Jackson, Frank, 8
Japan, 21, 121, 138, 148–9
Jonas, Hans, 33–4

Kalahari tribal society, 25
Kantorovich, Leonid, 116–17, 122
Kaplan, David, 16, 27
Kapp, Ernst, 13
Kay, John, 88
keyhole surgery, 101
Keynes, John Maynard, 32, 57, 144
Kline, Stephen, 12, 14–15
know-how, 62–3, 77, 115, 133, 143, 163, 177
'knowing that', 61–3, 76, 133, 163, 177
knowledge, 61–3
Korea, 21, 121

Korean piano industry, 157
Kuhn, Thomas, 67–75, 76, 78, 79

labour, 24–5, 32, 113
Ladyman, James, 78, 81
Lakatos, Imre, 71–2
Large Hadrom Collider, 53
Laudan, Larry, 73–4
laws, 143–4
Lazonick, William, 147–8
Lee, Keekok, 12, 16, 44
life expectancy, 102–3, 114
linear programming, 116–17
logic, 75–8, 133–4, 137
looms, 88

machine technologies, 84, 88
magnesium, 95
mainframe computers, 105–6
malaria, 101
management of technology, 151–62, 193–4
manufacturing technology, 193
Marcuse, Herbert, 32–3
market artefact, 118
market criterion, 95–6
market economies, 121–3, 127–9, 151–62
markets, 48, 96–7, 125–6
Marx, Karl, 7, 20–1, 25, 65, 66, 119, 132, 187, 192–3
mass communication, 137
mass production, 33, 84, 143
Masterman, Margaret, 71
material culture, 138–40
materials technology, 84
mathematics, 26, 48, 57–9, 78–9, 117, 118
McMullin, Ernan, 71–4
medical technology, 2, 3, 15, 100–3, 190
Meitner, Lise, 84
Mendeleev, Dmitry, 70
metal tubes, 87–8
metaphysical abstractions, 54
metaphysics, 5, 8, 48, 61, 108, 134
microscope, 19–20, 23
Microsoft, 96
military technology, 129–30, 188–9
mining technology, 84–5, 144–5, 185–6
modeling technology, 118–19
modernity, 3, 10, 11, 16, 28, 55, 71, 134, 137, 139
money economies, 121
Moreno, Roland, 126
multinational corporations, 131
music, 2, 196–7

natural resources, 2, 8
  human reconfiguration of, 44–5
  reconfiguration of, 44–8
nature, 4–5, 12, 26, 30, 45, 52–61
  definition of, 53
  exploitation of, 144–5
  humanity and, 61, 164
  infinity in, 59–60
  knowledge about, 61–3
  mathematics in, 57–8
  nature of, 60–1
  probability in, 55–7, 59–60
  protection from, 2
  purpose in, 59
  science and, 162–3
  technology and, 81
  time, force and field in, 54–5
Newcomen, Thomas, 94
Newton, Isaac, 54, 55, 134
Newtonian theory, 80
Niedrach, Leonard, 98
nitrogen, 91
Nobel, Alfred, 122
non steroid anti inflammatory drugs (NSAIDs), 101
nuclear fission, 84
nuclear power, 6, 17, 23, 39–40, 83–4, 85
nuclear waste, 109

oestrogens, 109
ontology of humanity, 201–3
optics technology, 19
outsourcing, 137–8

Paine, Thomas, 21
pain management, 3, 102
Palley, Thomas, 125
Papin, Denis, 94

paradigm shifts, 67–8
particle physics, 52–3
pattern recognition, 141
penicillin, 23
pensions, 124, 200
person/people response, to technology, 181–2
pesticides, 6, 90–1
petrochemicals, 108–9
Pfaffenberger, Bryan, 138–40
pharmaceuticals, 100–1
phenomenology, 35
philosophy of technology, 155
  academic literature on, 6–40
  issues for, 7–8
  systems network concept, 41–145
physical entities, 53
physicalism, 134
physical matter, 4–5
physical processes, 53–4
physics, 55
Pinch, Trevor, 7, 27
Planck, Max, 69, 79
plastic industry, 85
plastic waste, 108–9
Plato, 76
pollution, 108–9
Popper, Karl, 55, 57, 64–7, 68, 70, 75–80
population growth, 143
post-modernity, 3, 4, 11, 28, 137–9
power generation technologies, 85, 87, 109–10, 190–1
price, 122–3
price/performance criterion, 95–6, 126, 155
primitive societies, 130, 142
privatisation, 127–9
probability, 55–7, 59–60, 66
production, 3, 20–1, 32
  food, 90–3
  mass, 33, 84, 143
  outsourcing of, 137–8
production function, 113, 115
productivity, 49, 84, 89, 90, 112–15, 124–30, 137, 141–3, 184–7
productivity paradox, 106
propensity hypothesis, 65, 80
propulsion technologies, 93–8
pseudoscience, 65
public sector, 127–9, 189
purpose, in nature, 59

quantum mechanics, 12, 60, 141
quantum theory, 56, 78–9

rationality, 8–12, 18, 44, 55, 68, 69, 71, 74, 75, 78, 137
realism, 73, 75, 79–81
reason, 70–1, 74, 133–4, 137
reasonableness, 134
reduction, 75, 78–9
reductionism, 141
refining technologies, 85
refrigerator, 3
reification, 31–2
relativity, 65, 79
research and development (R&D), 148–54, 156–7, 158
Ricardo, David, 113
ritual, 139–40
Robert, Richard, 88
Romanticism, 11
Rosenberg, Nathan, 22–3, 157–8
Russia, 21–2, 121, 138, 149, 185–6

Salmon, Wesley, 77
Savery, Thomas, 94
savings algorithm, 117–18
Scharff, Robert, 29
Schoenbein, Christian, 98
Schumpeter, Joseph, 82, 146–7
science, 26, 27
  applied, 133
  economy and, 163
  humanity and, 162
  inductive method, 75–8
  Kuhn on, 67–75
  nature and, 162–3
  in network systems model, 61–81
  objectivity of, 71–2
  philosophy of, 64–7, 75, 80–1
  Popper on, 64–7
  realism and, 79–81
  reduction in, 78–9
  society and, 163
  technology and, 81, 82, 163

Science Policy Research Institute (SPRU), 146
scientific laws, 77–8
scientific paradigms, 67–75
scientific processes, reconfiguration of, 44–8
scientific revolution, 68, 70, 72
scientific theories, 65–75, 80, 133
Sears Roebuck, 143
selective catalytic reduction (SCR), 87, 110
semiconductors, 115
sewage treatment, 192
shadow prices, 122
Shapere, Dudley, 72
Shapiro, Stewart, 57–8
Siemens, 111–12
'Silent Spring' (Carson), 6, 90
Silicon Valley, 149
Simpson's paradox, 63
single event probability, 66
Skidelsky, Robert, 57
smallpox, 2, 101
smartcard, 126–7
Smil, Vaclav, 2, 83, 84, 114–15, 125
Smith, Adam, 116, 137
Smuts, Jan, 141
Snow, C. P., 28
social classes, 134
social constructivism, 27–8
socialism, 22, 119
social networking, 108, 131
social structures, 20–2, 25, 28, 49, 50, 130–44
society, 130–44, 163, 196–7
sociology, 140–1, 143
sociotechnical system, 139–40
soft determinism, 25
Sokal, Alan, 80–1
Solow, Robert, 115
South Korea, 138, 152
Soviet Union, 1, 116–21
Spence, Thomas, 21
spinning jenny, 88
spreadsheet software, 96
steam engine, 24, 84, 93, 94
stilbestrol, 91
Strassman, Fritz, 84
stress-related diseases, 3
string theory, 53
structuralism, 43
sub-atomic particles, 52–3, 60
sulphur dioxide, 109–10
sulphur trioxide, 109
Switzerland, 152
systems equilibrium, 50–1
systems network model, 42, 43, 49
  balance of power in, 169–72
  dynamism in, 50–1
  ecology in, 144–5
  economy in, 115–30
  entities in, 52–145
  implications of, 179–200
  interactions of, 162–9
  main direction of, 177–9
  model assumptions, 44–52
  nature in, 52–61
  productivity in, 112–15
  science in, 61–81
  society in, 130–44
  technology in, 81–112
systems technology, 104–8
Szilard, Leo, 83, 84

technocracy, 28
techno-humans, 15, 143
technology
  see also specific types of technology
  academic literature on philosophy of, 6–40
  analytics of, 18–37
  applied, 179
  as artefact, 201–3
  behavioural economics view of, 146–50
  business response to, 192–5
  in command economies, 120
  commercialisation of, 126–9
  competing, 95–6
  consumer response to, 182–4
  control of, 4
  definitions of, 8–18, 82–3
  development of, 23–4, 147–50
  education response to, 195–6
  government response to, 197–200
  humanity and, 81–2, 172–7
  as human reconfiguration of natural materials and processes, 44–8

technology – *continued*
impact of, 1–4
interaction between humanity and, 49–50
management of, 151–62, 193–4
market determination of, 95–7, 126, 158–62
moderating, 37–40
narrative, 83–8
nature and, 81
objective of, 155
person/people response to, 181–2
philosophy of. *see* philosophy of technology
as predictable, 51
productivity and, 141–3
science and, 81, 82, 163
society response to, 196–7
in systems network model, 81–112
understanding of, 4, 44
voter response to, 187–92
worker response to, 184–7
technology business audit process, 161–2
technology transfer, 158
telecommunications, 106–8
telephony, 16–17
television, 3, 137
testability, 66–7
textile technologies, 88–9
Thatcher, Margaret, 118
thermal efficiencies, 87
thermostats, 104
trains, 97, 127–9
transportation technology, 189–90
Trevithick, Richard, 94
Twitter, 108

uncertainty principle, 56
underground coal gasification (UCG), 85
United Kingdom, 152
British Rail, 127–9
textile industry in, 88–9
United States, 21
auto industry, 94–5
market economy, 121
R&D in, 152
Silicon Valley, 149
universities, 131
uranium, 84
urbanisation, 49, 143, 196
utopia, 29–34

value chains, 126, 156
van Fraasen, Bas, 79–80
Verbeek, Peter, 16, 35–6
VHS technology, 95–6
voter response, to technology, 187–92

Walton, Ernest, 83
waste disposal, 108–9, 192
Watt, James, 94
Webb, Sidney, 119
Weber, Max, 10
web sites, 108
White, Lynn, 145
Whittle, Frank, 97
Wilson, Harold, 118
wind power, 85
Winner, Langdon, 28, 34, 35
word processor software, 96
worker response, to technology, 184–7
World Wide Web (WWW), 108
Wright, Crispin, 57–8
Wright, J. W., 117

zero-emission electric power, 85
Zhiguli car, 120